第一次全面剖析建筑如何自然地植入城市

Total Analysis of How Architecture Naturally Integrates into the City

新印象 New Impression

中国建筑工业出版社
CHINA ARCHITECTURE & BUILDING PRESS

生长的文化建筑
Growing Cultural Architecture

昆山文化艺术中心全解析
Everything About Kunshan Culture & Art Center

策划&主编：于冰

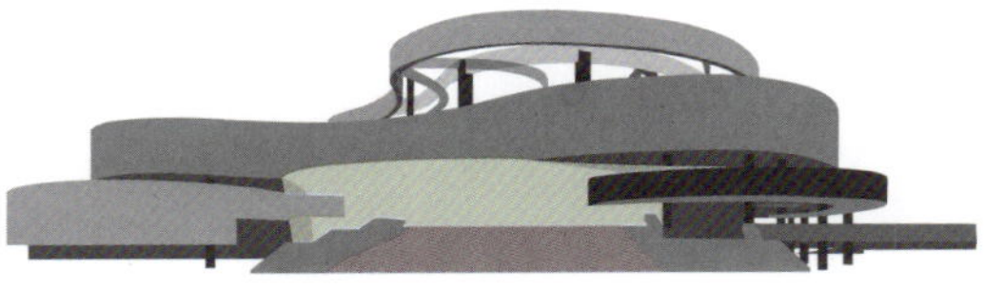

第一次全面剖析建筑如何自然地植入城市
Total Analysis of How Architecture Naturally Integrates into the City

文化·传承·思考 Culture and Thinking/ 理念·功能·剖析 Idea and Analysis/ 城市·设计·生活 Urbanization and Design

目录 CONTENTS

文化·传承·思考
Culture·Inheritance·Thinking

本章悉心梳理了昆山市厚重的人文历史、浓郁的地域特色、经典的文化传承等传统脉络，揭示了建筑师创意思维的由来；并通过采访来展现昆山的城市规划蓝图，勾勒出昆山规划与城市设计的思索之旅及其与建筑个体的相互关系。

This chapter carefully sorts out Kunshan's profound culture and history, strong geographical characteristics and classic cultural development and explores the origin of architects' creative ideas; and through interviews unfolds the Kunshan urban planning blueprint and depicts the relationship between the journey of Kunshan planning and urban design and the individual buildings.

采访：许振敏，曹勇 Interview with Xu Zhenmin, Cao Yong

理念·功能·剖析
Idea·Function·Analysis

为实现"房在园中"的构思意图、建筑造型特点及功能要求，甲方、设计、材料、施工各环节密切合作，在制作模型、实验、模拟，反复测试与论证的基础上，最终不仅完整地实现目标，并为业界提供了很多值得借鉴的设计与实践经验。

To achieve the design idea, architectural features and functional requirements of "room in the garden", architects worked closely with clients in terms of design, materials, construction and other aspects. On the basis of modeling, experiments, simulations, repeated testing and demonstration, they finally achieved the whole design objective, providing the industry with a lot of design and practical experience.

采访与文章：崔愷，阿兰·博尼，李跃文，金照彬，范晓玲 Interview & essay: Cui Kai, Alain Bony, Li Yuewen, Jin Zhaobin, Fan Xiaoling

城市·设计·生活
Urbanization·Design·Life

昆山规划以及文化艺术中心的建造，汲取了国外经典案例的规划思想。不仅给居民带来了福祉，并为昆曲等传统文化的再生提供了一个互动舞台，让建筑自然植入城市，选择了一条适合城市与其居住者们的道路。

Kunshan planning and the construction of Kunshan Culture and Art Center draw on the planning ideas of foreign classic cases, providing not only benefits to the residents but also a platform to Kunqu opera and other cultures to revive and interact with people, implanting buildings naturally into the city with a path suitable for both the city and its residents.

采访：周继春，姚伟宏，韩明 Interview with Zhou Jichun, Yao Weihong, Han Ming

序言 I
行云流水

为一个地方设计一座大剧院，就好像为这个地方咏唱一首赞美的歌。五年前，当我应邀为昆山设计大剧院时，心中自然而然萌生出这样一种情感，并以这种情感的冲动去看待这片土地，孕育它的未来。

位于阳澄湖畔的这个朝气勃勃的城市，它原本也是江南水乡。昆山隶属于苏州，紧邻大上海，堪称是江苏经济区的桥头堡，大量的台企入驻让这里的经济快速发展，使之成为全国百强县之首。而这些年昆山在生态治理上也狠下工夫，为了阳澄湖的水质而限养驰名的大闸蟹，为了城市绿肺，在市中心种下大片森林。只可惜，在快速城市化进程中，那些老街老巷消失了，城市文化也伴随着城市特色的衰落而日渐式微，建设大剧院就是这种背景下为了重振文化，追求特色，让昆曲的故乡文脉绵长的规划步骤之一。

这个重要的城市公共建筑在特定的环境中建设首先碰到的是布局问题，布局的策略不同会产生不同的城市空间关系，也影响建筑设计的方向。经过分析，我们选择了另一路径，把用地与北面的森林公园联系起来，当作连续的城市绿带公园来统一考虑，而将剧院作为公园里的建筑，与河道发生更密切的关系，强调它的滨水性，也试图建立与体育场馆更积极的空间联系，在尺度上、色彩上也应与之协调对话。

昆山艺术中心功能综合，技术复杂，完全是一个团队合作的成果。首先感谢昆山城市建设投资发展有限公司周继春董事长对设计师的尊重，项目总经理范晓玲对工作的认真负责使我们对工程的设计一直处于有效、有序的控制之中。在艺术中心设计施工的全过程中，业主表现出很好的职业素养和追求，不熟悉的就去考察学习，不了解的就找专业咨询，不放心的就反复做样板比较，并且每一个重要的技术性决定都要请建筑师来做，表现出了充分的信任，让我们的合作十分愉快，大家早就成为了好朋友。当工程结束范总调升城建局副局长，我既为她的进步而高兴又有些不舍，毕竟艺术中心还没收尾，整改的工作没完，也还有二期项目等待启动呢。事实上在昆山另一个文化项目中我们又与周总的团队开始新的合作了！

此外，应该感谢我们的设计团队，我们设计院建筑、结构、机电、总图、室内设计等各专业的设计师，为这个项目团结奋战付出很多，对工作的认真和解决问题的能力的确得到了业主和施工各方的认可和好评。另外要特别感谢与我们合作声学设计的张三明教授，他对不对称观众厅的设计挑战很有耐心和自信地应对，使观众厅的声学质量达到了较高的水平。值得一提的是，法国著名艺术设计师阿兰·博尼先生虽然进入这个项目较晚，但他们认真地倾听建筑师的构想，善于用彩色肌理和灯光去表达艺术的效果，并积极地与室内设计师合作，多次深入现场，选样板、调灯光，让建筑的艺术性有了很大的提高，得到了业主的认可。实际上，艺术设计在国内工程中还很少作为一个专项设计参与进来，而在国外却是较普遍的，这对于提高建筑的文化艺术品质十分必要。博尼先生先后参加的包括国家大剧院在内的许多国内外项目的成功建设充分说明了艺术设计的重要性。

最近又听说这个刚刚落成的建筑——文化艺术中心被市民选入苏州市十大优秀公共建筑，让我既很意外，也很欣慰。这个项目虽然完工了，但我对它仍报有一种期望，一个文化建筑应该自然地融入这个城市的生活。如同我每个作品完成后一样，期望过若干年后我回到那里，看到它跟城市生活真正地、密切地联系在一起了，这才是我最希望看到的。

崔愷
中国建筑设计研究院（集团）建筑设计总院名誉院长、集团副院长、总建筑师

Foreword I
Floating Clouds and Flowing Water

To design a theater for a place seems like composing a song of praise for the place. I was overwhelmed by such kind of emotion and felt the impulse to treat this land and nurture its future with such emotion five years ago when I was invited to design a theatre for Kunshan.

Located near Yangcheng Lake, this vibrant city used to be a watertown. Kunshan belongs to Suzhou and is near Shanghai, known as the bridgehead of Jiangsu economic zone. A host of Taiwan enterprises flood in and fasten the local economic development, making Kunshan the National Top 100 County. During these years, Kunshan has also put great efforts in ecological management. To improve the water quality in Yangcheng Lake, it has restricted the breeding of the famous hairy crabs; to freshen the urban green lung, it has planted a large stretches of forests in the urban center. It is such a pity that old lanes and alleys have vanished in the road of rapid urbanization and the urban culture has been casted away with declining urban characteristics. Under such background, to build a grand theatre is one of the planning steps to revive the culture, to pursue urban features and to make the birthplace of Kunqu opera long lasting.

To construct such an important urban public building in a particular environment, the first problem encountered is the layout, because different layout strategies will not only lead to different urban spatial relationship but also influence the direction of architectural design. After analysis, we chose another way to design. We unified the land with the Forest Park on the north and considered the park as a continuous greenbelt of the city and the theatre as a building in the Park, so that we linked the theatre more closely with the river and emphasized its location at riverside. We also tried to establish a more positive spatial relationship between the building and the Sports Center, blending the two architectures in harmony in both scale and color.

The comprehensive functions and complex techniques of Kunshan Culture and Art Center are entirely the outcome of teamwork. We should first thank Zhou Jichun, chairman of Kunshan City Construction Investment & Development Company for his respect for designers and general manager Fan Xiaoling for her responsible work, which enabled us to keep the project design effectively and orderly. During the whole process of design and construction of the Art Center, clients displayed outstanding professional quality and aspiration by investigating and learning the unfamiliar professional knowledge and making sample comparison for the unsure. Every important technical decision was made by the architect, showing the sufficient trust upon us and making the cooperation very pleasant. We have already become good friends. When the project was nearly finished, manager Fan Xiaoling was promoted as the vice-director of City Construction Bureau. Thought happy for her progress, I felt sad to see her leaving, since the Art Center had not yet been fully completed and the restoration work had not yet been done and the Phase II project was about to start. Actually, we began to work and cooperate with the team of chairman Zhou again in another cultural project of Kunshan.

We should also give thanks to our design team – designers of architecture, structure, electro mechanics, overall layout and interior and other professional designers in the institute – for their united efforts on this project. Their working attitude and problem-solving ability are recognized and praised by clients and the construction company. Special thanks should be given to Professor Zhang Sanming who cooperated with us in acoustic design and responded to design challenges of asymmetrical auditorium with patience and confidence, upgrading the acoustic quality of the auditorium to a higher level. It should also be mentioned that famous French art designer Alain Bony and his team engaged in the project in a later phase, but they listened to the ideas of architects seriously. They excelled at using color, texture and lighting to express artistic effect and actively cooperated with the interior designers and made many on-site studies to choose sample plates and adjust lighting, greatly improving the artistic effect of the building and receiving acclaim from the clients. In fact, it is rare for artistic design to engage in domestic projects as a special design team, but it is rather common in foreign countries, which is significant to improve the cultural and artistic quality of buildings. Bony has participated in many successful projects at home and aboard, including the National Grand Theatre, which fully illustrates the importance of art and design.

Recently, I am surprised and delighted to hear that this newly completed building – Kunshan Culture and Art Center was selected by citizens as Top 10 Outstanding Public Buildings in Suzhou. Although the project has already completed, I still cherish a wish upon it, hoping that such a cultural building shall naturally blend with the urban life. Just like every other projects I have completed before, I hope that when I come back several years later, I will see it has truly and closely linked with the urban life. That's what I most want to see.

Cui Kai
Honorary president, vice president and chief architect of Architecture Design
General Institute of China Architecture Design & Research Group

序言 Ⅱ
让建筑承载城市记忆

Foreword Ⅱ
Architecture Embodies the Memory of the City

我的家在昆山，1993年大学毕业到昆山开发区工作，从那时起就开始体会这座城市的飞速发展。曾经有个口号："昆山就是开发区，开发区就是昆山。""昆山之路"闻名全国，昆山的综合经济实力跃居全国百强县(市)之首，成为全国18个改革开放典型地区之一。

任何发展模式都是有代价的。我出生在农村，现在属于开发区。当每次开车经过小时候生活的村庄时，总想从现在笔直、宽阔的马路和简洁、现代的厂房间找到原来自己家的位置。梦里也几次回到村边那条学会游泳、经常钓鱼的河边。我明白了，不管搬多少次家，我心里的"家"一直在那里。

这种城市居住者与过去的一切之间失去相互依存的关系，就像某个人失去记忆一样，其后果不仅是严重的，在未来也是难以挽回的。一个城市在发展经济的同时必须尊重土地、尊重历史、保护传统文化，否则无法让居住在这里的人们真正认同它。

昆山文化艺术中心是为全体市民提供一个文化载体、一个可以共同参与的大舞台、一块可以传播艺术和繁衍传统的土壤，是让新、老昆山人从精神上产生对城市更多的认同感的地方。

它的选址坐落在昆山西部新区，作为昆山西部副中心的有益补充，有机地连接了北侧的森林公园与南侧的行政服务地块，形成了一个完整的、人性化的市民活动中心。这无疑将为西部副中心创造新的吸引力，使片区产生连续的、值得关注的热点，同时将很好地刺激这片区域的良性发展。更重要的是，它也适时地起到了以文化濡染昆山城市精神，在深厚的精神层面振兴城市发展的作用。

让建筑承载起城市的记忆，我们的设计团队，尤其是崔愷工作室的伙伴们，很好地诠释了这一诉求，他们将昆山本土文化中的精髓——水乡、昆曲与并蒂莲等元素很好地融入了建筑语汇，建筑和内装设计的统一使项目更加完善。为了使设计图纸更好地落地，崔院士几乎每个月都会亲临施工现场，与现场工作人员仔细沟通，确认每个样板，这样的态度令我感动，也鼓舞着大家的士气。项目的顺利实施还有赖于庞大的顾问团队的支撑，艺术、声学、舞台机械、灯光、音响、运营，以及施工、监理单位，每个单位都派出了最精干的人员。同时，我们业主团队也付出了艰辛的努力，现场负责人范晓玲在工地摔断肋骨后坚持留在施工现场协调各专业，直到数周后无法继续支撑才去医院就诊。参与项目的数千人，我们没有办法在工程上书写他们的姓名，甚至他们中的大部分人都没机会等到开业看一场演出就已经离开这座城市，但我永远感谢他们的努力和付出。

事实上在昆山，单就功能而言不缺好的建筑，但是让建筑呼应城市的文化与记忆，昆山文化艺术中心算是一个很好的尝试。我们希望它不光承载城市的记忆，更着眼于改变未来，为市民生活方式的转变提供可能，我们通过它表达对这片土地的热爱和对生活在这里的人们的尊重。

周继春

昆山城市建设投资发展有限公司董事长

Kunshan is my hometown. I have been working in Kunshan Development Zone since I was graduated in 1993; from then on I realized that the city was underway of rapid development. The comprehensive economic strength of Kunshan vaults into the first place of Top 100 counties (cities), being one of the national eighteen typical reform and open regions.

I was born in rural area where has become development zone now. I always want to find out my home among current upright, broad roads and modern factory workshops. I came back to the river near the village many times in my dream. I figured out that no matter where I move to, the home in my heart is always there.

The lack of relationship between urban residents and their past is not only serious, but also irretrievable. During the process of economic development, the land and its traditional culture of a place must been respected and protected, otherwise it will not be the identified authentically by people dwelling here.

Kunshan Culture and Art Center provides all citizens a cultural carrier, an involved platform and a piece of soil that can convey the art and cultivate the tradition, making it a place identified more spiritually. The site is located in new zone of western Kunshan and enjoys an organic connection of Forest Park to the north and administrative service zone to the southern; therefore it forms an integral and humanized activity center. It is no doubt that the project is an attractive hot spot of the western Kunshan's sub-center, and the impetus of the regional benign development, playing the function of spreading city spirit with its culture and promoting the development in deeper level.

To make architecture embody the memory of city, the design team, especially the partners of Cui Kai Studio, interpreted the demand appropriately. They took the marrow of local elements into the architectural language; the unity of building and interior design made the project complete. During the construction process, Cui Kai came to the site every month and checked out every detail carefully. For smoothly carrying out the project, it depended on the support of a large group of consultants, the elites from various fields. As clients, our team also made great efforts, general manager Fan Xiaoling insisted on working after broken her rib. I appreciate their efforts and contribution forever.

Kunshan Culture and Art Center is a successful try in the aspect of echoing the city culture and memory through architecture. We hope that it will not only undertake the memory of the city, also focus on the changes of future and supply the possibility of transforming people's life style. We express our devotion to the land and the respect to all people living here via this building.

Zhou Jichun

Board Chairman of Kunshan City Construction Investment & Development Co., Ltd.

第一章 Chapter 1

文化·传承·思考
Culture·Inheritance·Thinking

本章悉心梳理了昆山市厚重的人文历史、浓郁的地域特色、经典的文化传承等传统脉络，揭示了建筑师创意思维的由来；并通过采访来展现昆山的城市规划蓝图，勾勒出昆山规划与城市设计的思索之旅及其与建筑个体的相互关系。

This chapter carefully sorts out Kunshan's profound culture and history, strong geographical characteristics and classic cultural development and explores the origin of architects' creative ideas; and through interviews unfolds the Kunshan urban planning blueprint and depicts the relationship between the journey of Kunshan planning and urban design and the individual buildings.

采访：许振敏，曹勇 Interview with Xu Zhenmin, Cao Yong

理念·功能·剖析
Idea·Function·Analysis

为实现“房在园中”的构思意图、建筑造型特点及功能要求，甲方、设计、材料、施工各环节密切合作，在制作模型、实验、模拟，反复测试与论证的基础上，最终不仅完整地实现目标，并为业界提供了很多值得借鉴的设计与实践经验。

To achieve the design idea, architectural features and functional requirements of "room in the garden", architects worked closely with clients in terms of design, materials, construction and other aspects. On the basis of modeling, experiments, simulations, repeated testing and demonstration, they finally achieved the whole design objective, providing the industry with a lot of design and practical experience.

采访与文章：崔愷，阿兰·博尼，李跃文，金照彬，范晓玲 Interview & essay: Cui Kai, Alain Bony, Li Yuewen, Jin Zhaobin, Fan Xiaoling

城市·设计·生活
Urbanization·Design·Life

昆山规划以及文化艺术中心的建造，汲取了国外经典案例的规划思想。不仅给居民带来了福祉，并为昆曲等传统文化的再生提供了一个互动舞台，让建筑自然植入城市，选择了一条适合城市与其居住者们的道路。

Kunshan planning and the construction of Kunshan Culture and Art Center draw on the planning ideas of foreign classic cases, providing not only benefits to the residents but also a platform to Kunqu opera and other cultures to revive and interact with people, implanting buildings naturally into the city with a path suitable for both the city and its residents.

采访：周继春，姚伟宏，韩明 Interview with Zhou Jichun, Yao Weihong, Han Ming

1.Culture·Inheritance·Thinking

文化·传承·思考

在经历了经济迅猛发展之后，昆山迫切需要由新兴产业基地向现代化城市的全面转型升级，为此推出了新的昆山城市总体规划修编方案。虽然昆山因其近年来经济的高速增长而一跃成为中国大陆经济实力最强县级市，但是在规划进程中，这座城市却难能可贵地保持着相对冷静的思考，注重从悠远的历史文化传统中寻根溯源。陆续开展的城市功能项目没有标新立异，建筑风格也能很好地融入本土特色，甚至道路绿化选择的也是适合当地环境的树种。

而作为这一规划中有机组成部分之一的文化艺术中心，建筑师同样注重延续这个城市的传统文化底蕴与魅力，在传统与现代之间巧妙选择了另一条路径，其具有浪漫情怀的创意将昆曲水袖、并蒂莲，江南水乡的曲折进退等昆山文化精髓的韵味巧妙地融入这座现代造型建筑中。并通过开放的设计手法处理与周边环境、建筑的关系，恰到好处地弥补了原有区域的设计缺陷，起到了完善城市功能与肌理的织补作用。

本章也基于这样的背景而展开，悉心梳理了昆山市厚重的人文历史、浓郁的地域特色、经典的文化脉络，引领读者感受建筑师创意思维的由来。通过对亲历城市近年来改造进程的媒体与规划负责人的采访，展现了昆山规划蓝图，并探寻其在城市规划、城市设计的思索之旅及文化艺术中心在其中的地位、相互关系。

After the rapid economic development, Kunshan is in urgent need of full-scale transformation and improvement from an emerging industrial base to a modern city, thus putting forward a new revised Kunshan Urban Master Plan is necessary. As Kunshan's rapid economic growth in recent years, it has become the city with the strongest economic strength in China, it is commendable that in the planning process the city has remained relatively clear thinking and focused on the profound historical and cultural traditions where its roots and sources can be traced. Urban projects are being carried out one after another without doing anything unconventional or unorthodox. The architectural style is well merged with local characteristics, greening and landscape are well adapted to local environment.

Kunshan Culture and Art Center is an integral part of this plan. Architects emphasize on continuing the heritage and charm of traditional culture of this city, skillfully choose a path between the traditional and the modern, and innovatively integrate the romantic Kunqu sleeves, twin lotus, winding characteristics of watertown and other Kunshan cultural essences to the modern-style buildings. The design approach opens the building to its surrounding environment and context, which perfectly makes up the flaws of the original area and improves the urban function and mechanism.

This chapter therefore carefully sorts out Kunshan's profound culture and history, strong geographical characteristics and classic cultural development and leads readers to appreciate the origin of architects' creative ideas. Through interviews with the media that have witnessed the process of urban transformation in recent years and with the person in charge of the planning, the blueprint of Kunshan planning is revealed to explore its urban planning and urban design thinking and the status of the center in this city and their interrelationship.

淞南书场
淞南书场
146

顾炎武故居

昆曲发源地千灯

历史人文 History & Art

汲取历史人文艺术菁华

在如今昆山物质经济飞速发展，昆曲等精神遗产重新走向国际舞台之际，艺术中心的设计也恰植根于斯，生发于斯，继续展现昆山文化及其本身的新艺术魅力

昆山的城市历史和先贤遗迹在华夏文明中占有重要的一席之地。这座城市古名娄邑，春秋战国时期先属吴，后属越，继又归楚。秦时置娄县。南朝梁大同初年（535-536年）始名昆山县。唐天宝十年（751年），分置华亭县，昆山县治由昆山东城迁至马鞍山南。从古娄县到南宋嘉定年间，前后约有1500年，吴淞江沿岸为娄县（昆山县）管辖。这包括了今天上海的大部分地区。宋嘉定十年（1217年），分置嘉定县。元代曾升昆山为州，明代复为县。清雍正二年（1724年），分置新阳县，与昆山同城分治。民国元年（1912年），昆、新两县才并合为昆山县。

RICH HERITAGE FROM KUNSHAN'S HISTORY, CULTURE AND ART

IN TODAY'S KUNSHAN, KUNQU OPERA AND OTHER CULTURAL HERITAGES IS RE-ENTERING THE INTERNATIONAL STAGE. IT IS THE BASIS ON WHICH THE KUNSHAN CULTURE AND ART CENTER WAS DESIGNED AND BUILT

The long history of Kunshan city occupies an important position in the brilliant Chinese civilization with the city's historic development, ancient relics. Originally named Louyi, the city consecutively belonged to the Wu State, the Yue State and then to the Chu State during the Spring and Autumn Period and the Warring States Period. From the ancient Lou County of Qin Dynasty to the Jiading Period in the Southern Song Dynasty, the history spanned about 1,500 years. In the Yuan Dynasty, the county was upgraded to a prefecture, but in the Ming Dynasty it was reinstated as a county. By the second year of the Yongzheng Period (1724) in the Qing Dynasty, a Xinyang County was established separately and put under the governance of one city together with Kunshan. By the first year of the Minguo Period (1912), Kunshan County and Xinyang County merged as Kunshan County.

I HISTORICAL CHANGES 历史沿革

有史以来，昆山曾经有过几个别称，一是“娄城”。由于古娄江贯穿境内，娄又是二十八宿之一，秦始皇实行郡县制时，便有娄县。它最初属于会稽郡，直至东汉末才属于吴郡。古娄县的治所娄城，相传为吴王寿梦所筑，用以防备越国和长江北岸淮夷入侵。另一个别称是“鹿城”。相传吴王寿梦好猎，曾经在城西卜山下(今红峰新村)豢鹿狩猎。民国《江苏六一县志》载“昆山……别称鹿城，盖以……旧有西鹿城而名。又称玉峰。”然而很少有人知道昆山的另一个别称“武城”（或南武城）。武城位于阳澄湖的东北角，巴城镇境内。阳澄湖湖面辽阔，是吴国国都的天然屏障。吴王阖闾在湖的东、北、南三面筑城主要是防备从长江北岸渡江来侵的东夷。民间传说，武城是军事家孙武奉吴王阖闾之命修筑的。他运用阴阳八卦原理，结合湖畔地形地貌，精心构筑了一座“卦形”水寨军防城。占地约16km^{2}的武城，呈“出”字形，设十二处水道出口，充分利用了纵横曲折的河道所具有的虚实形势，运用八卦学说，构成一座充满神秘色彩的易守难攻的防御城。岁月流逝了二千余年，如今在阳澄湖东北的武城遗迹仍绰约可见。

昆山城墙的修建主要经历了从土城到砖城的过程。土城的修建者是元代知州费复初。元祐元年（1314年），昆山州治迁到太仓，至正十二年闰三月，浙江海寇方国珍攻占太仓，劫掠昆山。为了避免海寇的侵扰，知州费复初又把州治迁回昆山。元朝至正十七年（1357年），开始修筑土城，他带领军民修筑城池，东起城河，西起仓基河，南起正阳桥，北至马鞍山北麓，基本奠定了昆山城的规模。该城周长二千三百九十九丈，高一丈八尺；城外壕沟长二千三百五十九丈，深六丈，宽六丈；设有东、南、西、北、东南、西南六门。明朝弘治四年（1491年），知县杨子器在土城上建筑了城楼，东门叫宾曦，西门叫留晖，南门叫朝阳，北门叫拱辰，东南叫迎薰，西南叫丽泽。

昆山砖城的修筑是在明嘉靖年间。嘉靖五年（1526年），因为昆山濒临大海、寇盗多发等原因，副都御史周伦首先上疏皇帝，请求修筑砖城，之后，巡按御史陈蕙、巡抚欧阳铎也相继奏请改筑砖城，最后由顾鼎臣代替工部奏请朝廷将土城改为砖筑。后来，巡按御史陈蕙、巡抚欧阳铎、郡守王仪率先在昆山修筑了砖城。砖城在嘉靖十八年（1538年）二月开始建造，嘉靖十九年（1539年）五月完工，重新建造了宾曦、迎薰、朝阳、丽泽、留晖、拱辰等六个城门，增设了五座水关。该城周围二千三百八十七丈，城高二丈八尺，城宽一丈四尺，总共有四千五百八十七垛雉堞。城外壕沟还和原来一样。

昆山古城建于江南水乡，城内有多条河流通过。城墙的建造者既考虑到城池的防御功能又充分考虑到老百姓生活的便利。江南水乡经常使用船只运输物品，所以在建设旱城门的同时还附带设有水城门，也就是水关。旱城门供车马等出入，水城门则供船只进出。

“六城门五水关”的说法，是昆山建成砖城之后形成的。明嘉靖年间重新建造了六个城门，增设了五座水关。宾曦、迎薰、丽泽、留晖、拱辰五个城门是“水陆并举”。这五座水关与城内的多条河流相连，至和塘水从留晖门流到宾曦门，穿城而过，横贯东西，是城内的交通干道；山塘泾则南连至和塘，北通拱辰门；自至和塘向南有一条河流环绕整个县治，这条河叫市河，俗称玉带河；东城河从宾曦门附近的至和塘曲折南流到达迎薰门；筛谷浜则从留晖门附近的至和塘南流到达丽泽门。另外，丽泽门内的河流可以从舞雩桥通到陈老桥、莼菜桥。砖城的建造者就这样根据城内水道的分布，恰如其分地设计了六个旱门、五个水门，俗称“六城门五水关”。它奠定了昆山市的城市规模。

Throughout history, Kunshan has taken several names, and one was Lou Town. In view that the ancient Lou River ran through this territory and Lou is one of the 28 lunar mansions, Emperor Qinshihuang kept Lou County as early as he started to implement the system of prefectures and counties. It was originally a part of Kuaiji Prefecture until the late Eastern Han Dynasty when it belonged to Wu Prefecture. Lou Town, seat of the government of the ancient Lou County, was said to be built by Shoumeng, King of the Wu State, for the purpose of preventing invasion by the Yue State and ethnics on the northern bank of the Yangtze River. Another name of Kunshan was Lu Town (or Deer Town). According to legend, Shoumeng loved hunting and ever chased deer at the foot of Xibu Hill (now Hongfeng New Village). "Kunshan ... also known as Lu Town, maybe because... there ever had been a West Lu Town. It was also named Yufeng Mountain", recorded in *Annals of Jiangsu Sixty-one County*. But few people know another name of Kunshan, Wu Town (or South Wu Town). Wu Town was located in the northeast corner of the Yangcheng Lake in the Bacheng Town. Wide and vast, the Yangcheng Lake served as a natural barrier before the capital of the State. Helv, king of the state, built fortifications on the eastern, northern and southern sides of the lake primarily to protect against invasion by ethnics that crossed the north shore of the Yangtze River. In folklore, Wu Town was built by the military strategist Sun Wu at the order of the King. By applying principles of Yin, Yang and Bagua to the lake topography, Sun carefully constructed a Bagua-shaped fortress stationed with country seamen. Covering an area of about 16 square kilometers, Wu Town took the shape of character "出" (which means "out"). The twelve outbound waterways fully took advantage of the twists and turns of the river courses. Based on the Bagua doctrine, they built a mysterious town that was easy to defend but difficult to invade. And now, two millennia later, ruins of the town are still visible to the northeast of the Yangcheng Lake.

Walls of Kunshan were built mainly with earth, then bricks. The earth town was built by a magistrate in the Yuan Dynasty named Fei Fuchu. In the first year of the Yanyou Period (1314) of the Yuan Dynasty, the Kunshan government moved to Taicang; and in Intercalary March of the twelfth year of the Zhizheng Period, a Zhejiang pirate Fang Guozhen captured Taicang and looted Kunshan. In order to avoid intrusion of pirates, Fei Fuchu moved the government again back to Kunshan. In the seventeenth year of the same period (1357), the earth town began to be built. Fei led the army and civilians in the construction and formed the general framework of Kunshan Town which was surrounded by the moat in the east, the Cangji River in the west, Zhengyang Bridge in the south and northern foot of the Ma'anshan Mountain in the north. The town measured 2,399 zhang in perimeter, 1 zhang and 3 chi in height; outside the town, the trench was 2,359 zhang long, 6 zhang deep and 6 chi wide; and six gates were set in the east, south, west, north, southeast and southwest walls. By the fourth year of the Hongzhi Period (1491) of Ming Dynasty, the magistrate Yang Ziqi built towers on the earth walls, with the east gate called Binxi, west gate called Liuhui, south gate called Chaoyang, north gate called Gongchen, southeast gate called Yingxun, and southwest gate called Lize.

Construction of the brick town took place in the Jiajing Period. In the Jiajing fifth year (1526), in view that Kunshan was near the sea and piracy was frequent, vice censor-in-chief Zhou Lun first made a proposal to the emperor for the construction of a brick town. Then the move was followed by senior officials like Chen Hui and Ouyang Duo. Finally Gu Dingchen, on behalf of the Ministry of Works, petitioned the court to change the earth town to a brick one. Later, Chen Hui, Ouyang Duo and the prefecture magistrate Wang Yi took the lead to build the brick town. The work started in February of the Jiajing eighteenth years (1538) and was completed in May of the nineteenth year (1539), with the six gates re-built and five water passes added. The new town measured 2,387 zhang in perimeter, 2 zhang and 8 chi in height, and 1 zhang and 4 chi in wall thickness, with 4,587 battlements. The outside trench remained the same as the original.

The old town of Kunshan was built on the watery land of the Yangtze River Delta, with a number of rivers passing along. In building its walls, people took into account both the defensive function of the town and daily convenience of the people. In watery regions like this, transport was always realized with vessels, therefore at the same time of building non-water gates, water passes also needed to be built. The former were used for traveling of horses and carriages, while the latter were for boats.

The description of "six gates and five water passes" was formed after the brick town of Kunshan was completed. In the Jiajing Period, the six gates were re-built and five water passes were added. Among them, Binxi, Yingxun, Lize, Liuhui and Gongchen Gates were connected to the city's many rivers and used for both land and water transport. Water of Zhihe Pond flew through the town from Liuhui Gate in the east to Binxi Gate in the west, and the course served as the main road of the town; Shantangjing was connected southward to Zhihe Pond and northward to Gongchen Gate; to the south of Zhihe Pond a river ran around the entire county government and its name was Town River, commonly known as Yudai River; the east moat twisted from Zhihe Pond near Binxi Gate southward till Yingxun Gate; Shaigubang River flew from Zhihe Pond near Liuhui Gate southward to Lize Gate. In addition, rivers inside Lize Gate could flow from Wuyu Bridge to Chenlao Bridge and Chuncai Bridge. In that way, builders of the brick town appropriately designed the six non-water gates and five water passes in light of the distribution of the waterways which constituted the contour and framework of Kunshan City.

开篇：建于明代的十眼长桥有九柱十孔，全长52m，造型古朴别致，为远近水乡所罕见，人称“小宝带桥”。
对页：锦溪镇老街菱荡湾旁的回廊。
本页，从左到右：天水桥俗称北观音桥。明代永乐5年由郭子敬捐资修建。历经600多年的风霜雪雨，仍保持着原有的恬静风貌。锦溪像中和双桥这样的回廊很多,让人至今还可在雨天中过街穿巷而不受天气的影响，不仅承载的是历史，也体现出前人的人文智慧。

Opening page: The ten-eye long bridge with the length of 52 meters has nine pillars and ten arches; the model is simple and novel, and was rare in neighbor region of lake and river, so called "Little Ribbon Bridge".
Opposite page: The winding corridor is beside Lingdang cove on Old Street in Jinxi Town.
This page, from left to right: Tianshui Bridge as North Guanyin Bridge has been built with the donation fron Gu Ziyi in Yongle period of Ming Dynasty. Through about 600 years, it is still standing gracefully as before. A lot of bridges with winding corridor exist in Jinxi. This design brought the convenience that people can cross the streets and alleys without the influence of weather. It not only bears the historical remains, but also represents the humane intelligence.

II RELIGIOUS MONUMENTS 宗教古迹

昆山还是一个历史上深受佛教道教濡染的城市，许多古寺分布其间，如马鞍山路北的华藏寺、阳澄湖边的崇宁古寺，千灯镇延福禅寺，锦溪镇莲池禅院，周庄的澄虚道院等。其中以延福禅寺，莲池禅院最为著名。

延福禅寺作为昆山市著名古刹之一，梁天监二年由千灯镇人王束舍宅捐建，取名延福禅寺。吴越天福二年（937年）敕赐"波若寺"，宋大中祥符元年复名为延福禅寺。曾做过县主簿的王珏，因年老无嗣，舍宅扩建寺院，该佛寺规模宏伟，是佛家圣地，江南民众前来朝拜，香火不绝。元末寺毁。明洪武年间（1368－1398年）又重建起。永乐二年，户部尚书夏原吉浚治吴淞江及千灯浦时，曾借宿于此寺。万历年间（1573－1619年），寺僧瞻月重修法堂、山门。明末寺毁，清寺僧大野密音重建佛殿、经阁。清同治年间，清军与太平军作战时，该寺毁于战火，仅存后房数间及和尚数名。解放后和尚还俗，房屋移属医院。2003年～2005年期间，政府搬迁了医院，由佛教协会重修恢复原貌。

莲池禅院则是"锦溪八景"中最负盛名的景点之一。相传宋孝宗陈妃葬于五保湖中后，建寺于傍近水冢的菱荡湾南侧，派僧人诵经护墓。疏池栽莲，荷香四溢，故寺名莲池禅院，又名古莲寺、古莲池。因寺内有天君王灵官的塑像，被俗称为灵官殿。莲池禅院历经宋、元、明三代，至清顺治十三年（1656年），若山和尚来此住持，着手扩建鼎新。里人孙垣捐银五百两，使殿堂、廊庑，次第告成，计有弥勒、三世、大悲、观间等殿，傍添斗坛、方丈、三官堂、鹤来轩等处。寺外筑长堤开巨沼，前后榆柳婀娜有致，放生池里芙蕖风摆掩映，别有韵味。乾隆三十二年（1767年）六月初五，一场特大风灾袭劫锦溪镇，房倒屋坍，船覆树飞，人畜伤亡，损失惨重。莲池禅院的殿宇、客堂、山门，全被飓风刮得片瓦无存。事后，和尚永辉及里人陆诞英等分别重建，大致恢复原貌。 直到乾隆三十八年（1773年）重建神道院文昌阁，将阁移建于莲池禅院。于是这里更为频繁热闹。1964年2月，莲池禅院改建为中学校舍，水乡泽国近千年的佛门之地，成了莘莘学子提高造诣之所。文昌阁下，书声朗朗，充满了蓬蓬勃勃的朝气。1993年锦溪中学迁至镇北新区，莲池禅院得以恢复。

本页，从上到下：千灯的延福禅寺作为昆山市著名古刹之一规模宏伟，包括天王殿、玉佛殿、大雄宝殿，以及著名的千灯镇标志建筑——秦峰塔，现址为2003年至2005年期间由佛教协会重修恢复的原貌；延福禅寺的砖雕照壁。

对页：多数庙宇寺院的大门总是向南开着的，而锦溪的莲池禅院却是大门朝北，这在中国的寺庙中，是绝无仅有的一例。于是就有了 "水乡佛国，庙门朝北，天下世界，只此一家"的说法。

This page, from above to below: As one of the famous ancient temples, the grand Yanfu Temple in Qiandeng includes Hall of Heavenly King, Jade Buddha Hall, Mahavira Hall and the well-known landmark building in Qiandeng town – Qinfeng Tower, and its current site is restored by Buddhism Association from 2003 to 2005; the brick carving on Yanfu Temple's screen wall.

Opposite page: Most of the gates of temple face to south, but the gate of Lianchi Temple in Jinxi faces to north which is the unique one among all of the temples. Therefore here is a saying "Buddhist temple in regions of watertowns, and the gate faces to north, all over the world just one example."

Kunshan is a city deeply influenced by Buddhism and Taoism in its history. Many temples are distributed in the city, such as Huazang Temple to the north of Ma' anshan Road, age-old Chongning Temple nearby the Yangcheng Lake, Yanfu Buddhist Temple in Qiandeng Town, Lianhua Buddhist Temple in Jinxi Town, Chengxu Daoist Temple in Zhouzhuang, and so on. Among them, Yanfu Buddhist Temple and Lianhua Buddhist Temple are the most famous.

Yanfu Temple, one of the famous, ancient temples of Kunshan City, was built with the donation of Wang Shu, a resident of Qiandeng Town who sold his house, in the second year of the Liang Tianjian Period. In the second year of the Wu-Yue Tianfu Period (937), the emperor renamed it Boruo Temple, and it restored the present name in the first year of the Dazhong Xiangfu Period of the Song Dynasty. Wang Jue, who ever served as the county registrar and had no offspring, sold his house at his old age to extend the temple. With a large size, the temple became a Buddhist shrine that continuously attracted crowds of nearby people to worship. In the Yuan Dynasty, the temple was destroyed. In the Hongwu Period (1368-1398) of the Ming Dynasty, it was rebuilt. In the second year of the Yongle Period, minister of revenue Xia Yuanji ever stayed for accommodation when he carried out water control in the Wusong River and the Qiandengpu River. In the Wanli Period (1573-1619), a monk named Zhanyue renovated the ritual hall and the gate. By the end of the Ming Dynasty, the temple was destroyed again, but a monk of the following Qing Dynasty named Daye Miyin reconstructed the Buddha hall and the scripture depositary. During the Tongzhi Period, the temple was destroyed in the war between Qing army and Taiping Army, with only several rooms and monks left. After the liberation of New China, the monks returned to secular life, and the houses became part of a hospital. From 2003 to 2005, the government moved the hospital and the Buddhist association rebuilt and reinstated the temple.

Lianchi Temple is in one of the most prestigious attractions among the "eight scenic spots of Jinxi". According to legend, after Concubine Chen of Emperor Xiaozong of the Song Dynasty was buried by the Wubao Lake, a temple was built on the south side of the Lingdang Bay that was close to the waterfront of the mausoleum. And monks were sent there to chant Buddhist prayers and guard it. With overflowing fragrance, the pond was sparsely decorated with lotuses, hence the temple got its name Lianchi Buddhist Temple (namely Lotus Pond Buddhist Temple), and it is also known as Gulian Temple or Gulianchi Pond. Since there is a sculpture of a god named Wagnlingguan, who is red-haired, dark-bearded, red-faced, glaring with three eyes, holding a pagoda and dressed in armor, the temple is called Lingguan Hall by the folk. The Temple survived the Song, Yuan, Ming Dynasties and began to be renovated and extended when a monk named Ruoshan came to serve as the abbot in the thirteenth year of the Shunzhi Period (1656) of the Qing Dynasty. With the donation of five hundred liang of silver by a countryman named Sun Yuan, the halls and verandas were sequentially completed, and then there were halls called Maitreya, Sanshi, Dabei, Guanjian, etc., with such scenic sites as Doutan Altar, Abbot's Room, Sanguan Hall and Helaixuan Room. Out of the temple, long causeways and a large pond were built, with elms, willows and other trees gracefully standing around. By June Fifth in the thirty-second year of the Qianlong Period (1767), a heavy storm hit Jinxi Town and caused great casualties and losses by destroying houses, boats, vegetation, and killing people and livestock. In the temple, halls, rooms and gates were all ruined completely by hurricane. In the aftermath, a monk named Yonghui and a countryman named Lu Danying, together with others, reconstructed the facilities respectively and roughly restored the original appearance. Six years later (1773), the Temple became more bustling and crowded as Wenchang Building was reconstructed and moved there. In February 1964, the temple was converted into a secondary school, thus this nearly one-millennium-old Buddhist shrine on the watery land became an educating place for numerous students. The sound of loud reading echoed near Wenchang Building and brought youthful vigor. In 1993, Jinxi School moved to Zhenbei New District, and the Lianchi Buddhist Temple was restored.

III A CULTURAL TREASURE-HOUSE 人文殿堂

昆山不仅有着悠久的历史，更是人文荟萃之地，其间留下了众多值得流连玩味的历史遗迹，被称为南宋“中兴四大诗人”之一的范成大，他留下的两首东禅廊夜二绝均写夜景。范成大经常读书至深夜，才可能对荐严寺的夜色有如此真切的感受和细致入微的描写。诗中所提到的东禅寺，古名为荐严资福禅寺。原来位于集街，后在朝阳路东重建。当时即居城之东隅，“东禅寺”之名由此而来。

在千灯镇，除了古老的特色民居，这里更有著名昆曲昆山腔的创始人顾坚的纪念馆。前来参加昆曲聚会的人用自己钟爱的方式来演绎着昆曲、传播着昆曲、享受着昆曲；每天纪念馆里来来往往、络绎不绝的游人，也在其清新而浓郁的地方风情中认识了昆曲、了解了昆曲、熟悉了昆曲。

熟悉史实的人们，来千灯镇更不会忘记瞻仰与王宗羲、王夫之并称为明末清初三大儒的顾炎武故居。顾炎武不仅在清初文坛上享有盛誉，并在思想上提出“行己有耻”，提倡经世致用。谈到“亡国”与“亡天下”的区别，顾炎武在其最有价值的代表作《日知录》中颇有见地指出，维护一个王朝的政权，是君臣等上层统治者的事，而避免整个民族的道德沦丧，则是每一个老百姓的共同责任。后来，他的这段话被凝练成广为人知的“天下兴亡，匹夫有责”八个字，对国人的精神境界影响深远。

有着重视教育文化传统的昆山，其私家藏书活动在历朝历代也始终有着极好的口碑。藏书传统对昆山人也产生了深远影响。尤其到了明清时期，昆山的传统文化呈现繁荣的景象，很多人著述等身，涌现出不少文化望族。家族之内，互相唱和，互相影响，成为一种风俗。昆山最著名的藏书楼有协助于谦抗击外虏的叶盛后人所建的“菉竹堂”，清康熙朝重臣昆山人徐乾学所建的“传是楼”等。

近年来，周庄的双桥故事已为人所熟知，上海著名旅美画家陈逸飞以双桥为背景，创作的名为《故乡的回忆》的油画，后被阿曼德·哈默用高价购藏。哈默先生在访问中国的时候，将这幅油画又送给了邓小平。因为双桥这段故事，周庄古镇声名鹊起。双桥是位于周庄中心位置的世德和永安两桥，始建于明代，桥面一横一竖，桥洞一方一圆，样子很像是古时候人们使用的钥匙，所以当地人称之为“钥匙桥”。陈逸飞的画使默默无闻的双桥走向了世界。酷似钥匙的双桥也因此成为了开启周庄与国际交往之门的钥匙。而有着“三十六顶桥，七十二只窑”之称的锦溪最有名的古桥有8座。它们是天水桥、太平桥、十眼桥、锦溪中和双桥、众安桥、溥济桥、普庆桥、里和桥。每个古桥都有着自身古老而独特的历史。

此外，昆山还走出了明代名医王履，著名的航海家费信，哲学家魏校，清代著名理学家、教育家朱柏庐，以及汉语文字改革的先驱朱文熊，土木工程师、工商界爱国人士俞楚白等名人先贤；在近现代的人物中，从小生活在昆山的美国电脑大王王安，更是以自己的聪明智慧，成为华人在海外创业的成功典范。城市的文化积淀和人才辈出的人文史，给今天的昆山留下了璀璨的记忆和精神遗产。至今，在这座城市的人文特质中仍显示着传统文化底蕴深厚的特点。

昆山文化艺术中心恰恰在延续这种印记的基础上，汲取了传统文化的滋养与现代昆山人开拓创新的意识，将昆山独有的城市气质与精神更充分地加以体现。

对页：在千灯镇口，两河交汇、三桥联袂，方泾桥和恒升桥和鼋渡泾桥组成的“三桥邀月”是千灯的经典人文景观之一。图为三桥中的两座桥，前面为明代的方泾桥，后面的三孔石拱桥是清代的恒升桥。

本页上图：千灯顾坚纪念馆是清代末民国初建筑风格的庭院。详细叙述了昆山腔创始人及同仁创造“昆山腔”的史实，也概述了中国戏曲与昆曲的发展历史。

左图，从上到下：顾炎武故居又名顾氏南宅“贻安堂”。南宅现存五进明清建筑，雕梁画栋，砖刻门楼十分精细，具有极高的艺术价值。千灯恒升桥细部。

下图：恒升桥下的大河原名为千灯浦，后来为纪念疏拓千灯浦的明代户部尚书夏元吉而改名为尚书浦。

Opposite page: At the entrance of Qiandeng Town, two rivers join together, three bridges as Fangjing Bridge, Hengsheng Bridge and Yuandu Bridge make a classic humane sight in Qiandeng named "Three bridges embracing the moon". View of two of the three bridges, the front one is Fangjing Bridge of Ming Dynasty, the back one is Hengsheng Bridge of Qing Dynasty.

This page, above: The memorial hall of Gu Jian is a garden with the architecture style in the transition period between Qing and The Republic of China. It also includes the sketch about the development history of Chinese Opera and Kunqu.

Left, from above to below: The Gu Yanwu's former residence also named Gu's south house "Yian Tang". The south house still maintained five architectures in Ming and Qing style, carvings on pillars and beams, bricks and doors are exquisite, enjoying a extraordinary high artistic value. The detailed section of Hengsheng Bridge.

Below: The river beneath Hengsheng Bridge was Qiandeng River, in order to memerize the Shangshu in Hu department Xia Yuanji who dredged the Qiandeng River, it was then changed name to Shangshu River.

Kunshan has both a long history and rich cultural resources. Many interesting historical sites have remained on this land. Fan Chengda, recognized as one of the "four greatest Zhongxing poets" in the Southern Song Dynasty, left behind two poems titled Night at the Corridor of Dongchan Temple, both of which described night sceneries. Fan often read late into the night, that gave him a real feeling and detailed idea on how to depict the night at the Jianyan Temple. The Dongchan Temple in the poems was called Jianyan Zifu Temple in the ancient times. Originally located on Jijie Street, the temple was rebuilt afterward on Chaoyang Road East. At that time it stood at the east corner of the city, hence it got the name Dongchan Temple.

In Qiandeng Town, there is the memorial of Gu Jian, founder of the famous Kunshan tone of Kunqu opera, in addition to the characteristic old houses. People who come to attend opera parties are interpreting, spreading and enjoying Kunqu opera in their favorite ways; the daily crowds of visitors to the memorial are also understanding, learning about and familiarizing themselves with the opera in a fresh but strong atmosphere that is unique to this locality.

People with certain understanding of the history would certainly not forget to pay tribute to the former residence of Gu Yanwu, who was ranked among the three greatest scholars of the late Ming and early Qing period, together with Huang Zongxi and Wang Fuzhi. Gu Yanwu won great reputation in the literary arena of the early Qing Dynasty, and he also put forward the idea of "He who in his conduct maintains a sense of shame" to advocate the application of learning to the practical world. Speaking of the distinction between "ending the country" and "ending the world", Gu eloquently pointed out in its most valuable masterpiece *Records of Daily Insight* that the regime of a dynasty is the business of the monarch, royal officials and other higher-ranking rulers, but defending the entire nation against moral decay is the common responsibility of every people. Afterwards, these words of him were condensed into a well-known saying: "The rise and fall of the world is everyone's responsibility", which has had a far-reaching influence on people's spirit around the county.

In Kunshan, there is a tradition of valuing education and culture, and private collection of books in this region won excellent reputation throughout all the dynasties. This tradition also had a profound impact on Kunshan people. In particular, in the Ming and Qing Dynasties, traditional Kunshan culture became prospering, many people engaged in writing books and articles and there emerged a large number of culturally notable families. Within these families, all the members read together and influenced each other, which gradually become a custom. The most famous ancient libraries of Kunshan include Luzhu Hall built by descendants of Ye Sheng who assisted Yu Qian in fighting against foreign invaders, and Chuanshi Building constructed by Xu Qianxue, an important official with a Kunshan origin who served the Qing Kangxi Dynasty.

In recent years, the story of the Zhouzhuang Double Bridges is becoming well known. Chen Yifei, the famous Shanghai painter living in the U.S., created an oil painting titled "Hometown Memories" on the basis of the famous Double Bridges, and it was bought by Armand Hammer at a high price. During his visit to China, Mr. Hammer gave this painting to Deng Xiaoping. Because of this story, the old town of Zhouzhuang won great fame. The two bridges are Shide Bridge and Yong' an Brideg in the center of Zhouzhuang that were originally built in the Ming Dynasty. One of the two features a horizontal deck and a square hole, while the other has a perpendicular deck and a round hole. They look much like a key used in the ancient times, so they were called "Key Bridges" locally. Chen Yifei's painting silently brought the Double Bridges before the world, and thus the key-like bridges became the key that opened the door of Zhouzhuang to international contacts. Jinxi is well-known for its "thirty-six bridges, seventy-two kilns", and among them, eight bridges are the most famous. They are Tianshui Bridge, Taiping Bridge, Shiyan Bridge, Jinxi Zhonghe Double Bridges, Zhong'an Bridge, Puji Bridge, Puqing Bridge and Lihe Bridge. Each of them has its own unique, long history.

In addition, Kunshan is also the hometown of a range of sages and historical figures including: Wang Lv, the famous doctor in the Ming Dynast; famous navigator Fei Xin; philosopher Wei Xiao; Zhu Bolu, the famous thinker and educator in the Qing Dynasty; Zhu Wenxiong, pioneer of Chinese language reform; Yu Chubai, civil engineer and patriot in the commercial community; and so on. Among the modern and contemporary figures, Wang Wang'an, the U.S. computer magnate who had grown up in Kunshan, completed a success story of a overseas Chinese in the business world all by his wisdom. The splendid culture and the long history highlighted by a number of influential figures have left valuable memories and spiritual heritage to today's Kunshan, and the deeply rooted traditional culture still shines in the civilized life of the city.

The Kunshan Culture and Arts Center, while continuing those historical traits, absorbs nutrients from traditional culture and the innovative, pioneering spirit of modern Kunshan people. It embodies Kunshan City's unique character and spirit in a better manner.

地域风格 Vernacular feature

水乡地域的传统内涵与延续

江南水乡在独特的地理环境中，因势利导，从而创造了自己独特的生存方式，同时也以传统智慧积极影响着现代设计

昆山的水乡古镇是在漫漫岁月中，因集散四乡的物资而渐渐形成的。由“断竹，续竹，飞土，逐肉”开始，由于多余和短缺，开始有了交易，有了朝聚井汲，有了市井，街道也就不断延伸。古镇总是由湖荡淀泖围拥着。人们依水为市，贴水而居。一字形、井字形或龙字形的河流贯穿其中。街上的交易和河里的水市组成了古镇的喧腾。青瓦粉墙的人家，后窗临河，前门面街，一根晾衣裳竹竿，就能把街两边的楼窗连接起来。一幢幢深宅大院，莫道墙垣坍圮，苍颜斑驳，耸峙的封火墙和高啄飞檐，却永远保留着阀阅人家的气概。

“东南财赋地，江浙人文薮”，正是旖旎的水乡风光、安定的生活环境、良好的文化氛围，吸引了许多文人雅士来到这里。他们在昆山的周庄、锦溪等江南古镇留下了无数历史遗迹，也使古镇形成了稻米莲歌、耕桑读律、科名相继的传统。这种崇文重教的传统，一直延续到了今天……

TRADITIONAL CONNOTATION AND INHERITANCE OF ARCHITECTURE IN WATER-TOWNS

WATER-TOWNS HAVE PRODUCED A UNIQUE LIVING STYLE BY MAKING USE OF ADVANTAGES OF THE GEOGRAPHIC CONDITIONS

The ancient water-towns in Kunshan gradually come into being during the process of acting as collecting and distributing centers for goods from various places. Starting from the time when people started to exist as a result of surplus and shortage, people gathered around the wells in the morning, markets took shape, and streets continuously extended. The water-towns are always surrounded by lakes and rivers. Grand courtyards, though with dilapidated walls and mottled color, together with the towering firewalls and protruding penthouses, still sustain the lofty quality of the noble and dignified families. The charming water-town scenery, peaceful living circumstances and sound culture atmosphere attracted lots of scholars and figures to come here and they further left numerous historical heritages in the ancient towns in Kunshan. At the same time, the ancient towns also formed the tradition of planting rice and lotus, breeding silkworm, learning laws and reaching top ranks in imperial examinations. This tradition of appreciation for literature and education has been sustaining till today.

开篇：长达6km的沿河石驳岸。锦溪古镇布局亦以河道为骨架，街道依水而建，房屋临水而筑，水、路、桥、民居、街市巧妙而自然地融为一体。
左图，从左到右：周庄古镇因势利导，依水筑屋，以水成路，因水建镇，从而创造了自己独特的生存方式；周庄小巷中具有江南特色的过街骑楼。
Opening page: The revetment along river reaches a length of 6 miles. The outlay of old town in Jinxi considered river way as framework, the roads constructed along rivers, the houses built close to rivers, all of these rivers、roads、bridges、residence、markets become an integral part with nature.
Left, from left to right: The old town guides the layout in the light of its general trend, where houses adhere to river, riverways are considered as roads and the planning of town is made because of the water route, thus the town creates its own unique way of existence; The sotto portico crossing the street with characteristics of south of the Yangtze River in alley.

I

ZHOUZHUANG 周庄

"中国第一水乡"周庄古镇自古以来便是"湖荡环列，咫尺往来，皆须舟楫"的地理形势。"贞丰"是周庄的原名，汉初开始这里一直叫"贞丰里"。"里"是古时的一个行政单位，以五十户人家为一里。北宋元祐元年（1086年），周迪功郎舍宅为寺，当地人都感其恩德，于是把"贞丰里"改名为周庄。在三湖一江的环抱中，整个周庄古镇仿佛水泊中的一个岛屿，占地面积仅0.47km^2，却呈现出江南水乡古镇"小桥、流水、人家"的格局与风貌。

周庄古镇在其九百多年的发展过程中，依照自然地理形势，人工规划、适当开凿，逐步形成了二竖二横的"井"字型水系格局。南北走向为竖的两条平行河道，一为富安桥河（现周庄旅游地图称之"南北市河"，按清代《周庄镇志》所载，称之为"富安桥河"），一为油车漾，其中富安桥河最为重要，在古镇中间穿过，北接急水港主航道，南连南湖，是古镇与外界联系的交通动脉，同时将急水港的活水源源不断地引进来，流向南湖，带动整个水系，形成循环活水，保持优良水质。东西走向为横的两条平行河道，一称后港河，一称中市河，除了给古镇所有居民带来最大的舟楫便利外，这两条小河同样为古镇内的整个水系增添了动力，得以保持优良水质。

周庄古镇的源头可追溯到5000年前的良渚文化时期，镇郊的太师淀，曾是新石器时期原始先民的聚居地。在独特的水乡地理环境中，古镇因势利导，依水筑屋，以水成路，因水建镇，从而创造了自己独特的生存方式。

在科学合理、充满诗情画意的水系格局中，民居与桥梁的组合风格，丰富多变。家家枕河，户户依水，水为路，桥也为路，古街、古巷、石拱桥、石梁桥、石驳岸、石码头、石栏、桥楼、水弄、水阁等传统建筑和谐地联成一体，以至于像水晶筷子似的小河箸泾也能衬托出"轿从门前进，船从家中过"的张厅奇观。

在周庄的近千户民居中，明清和民国时期的建筑至今仍保存有百分之六十以上，其中有近百座古宅院第和六十多个砖雕门楼，还有一些过街骑楼和水墙门，这在江南水乡是堪称典型的。在这些建筑中，最具有代表性的当数沈厅。沈厅位于富安桥东堍南侧的南市街上，坐北朝南，七进五门楼，大小房屋共有一百多间房屋，分布在约一百米长的中轴线两旁，占地两千多平方米，为江苏省重点文物保护单位。

沈厅原名敬业堂，清末改为松茂堂，由沈万三后裔沈本仁于清乾隆七年(公元1742年)建成。沈厅共有三部分组成，前部是水墙门和河埠，专门供家人停靠船只、洗涤衣物之用，为江南水乡的特有建筑；中部是墙门楼、茶厅、正厅，是接送宾客、办理婚丧大事和议事的地方；后部是大堂楼、小堂楼和后厅屋，为生活起居之处。整个厅堂是典型的"前厅后堂"建筑格局。前后楼屋之间均由过街楼和过道阁连接，形成一个环通的走马楼，为同类建筑物所罕见。而大堂楼木梁架造型浑厚，一律为明式圆形图案。地板大多是60cm左右宽的单幅松板，坚固结实，可以想见当年建造时工程的艰重。而大堂楼的栏杆与槅窗制作较为精致，与前厅的建筑风格有所不同，属徽派风格。位于周庄北市街上的张厅，则是周庄明代民居中保存最为完好的建筑之一。据《周庄镇志》记载，张厅为明代正统年间中山王徐达之弟徐孟清的后裔所建，名为怡顺堂。清初徐家衰落，怡顺堂为张姓所购。因门前有两棵玉兰树，春天玉兰花开时，燕子在屋檐下筑巢繁衍，张家便将怡顺堂改为玉燕堂，俗称张厅。张厅占地面积1884m^2，临河临街，前后五进，共有房屋70间，整个建筑为典型的前厅后堂格局，设计巧妙，雕刻精细，古朴雅典。走进门厅，第四进是楼厅，也称大堂楼，为当年宅主家人的生活起居之处。该厅具有明代建筑的典型特征，厅下布有一个个置于粗大厅柱之下的木础。历经五百多年沧桑，至今仍完好无损。其中的奥妙在于木础的制作自有一套严格的操作规程，先用楠木等坚硬的木材雕成粗坯，尔后，沉入煎沸的上好白油之中，待其吸足白油后，再取出来加工成形，因而经久不坏。楼厅里那副出自吴冠中先生之手的对联："轿从门前进，船自家中过"，着实让人把玩生味。大厅的东侧，有一条幽暗深长的"陪弄"。旧时，没有大事轻易不开正门，每逢婚丧喜庆或有贵宾来访，才打开大门，抬进轿子。平时家人进出都走陪弄。如今，陪弄成了别有风味的旅游通道，在幽暗的光线中不知深浅地向前走，一片含有绿意的光亮投来，眼睛顿时一亮：后院花木扶疏，春色如画，一条晶莹的小河奇妙地闪现在弄底，贴着墙根流来，又穿越水阁而去，堪称设计中的神来之笔。除了张厅，周庄的明代建筑还有迮厅，它位于富安桥北侧，为明代崇祯年间吴江望族迮文焕所建，前后共五进，现存第一、四、五进，其余已毁。主厅又称大书房，据《迮氏家谱》记载，清光绪年间，陆润庠在此受蒙从师，成年后考中状元，遂将当年念书的大厅称之为大书房。现在大书房已不复存在，只有宽阔宅基上残留的块块方砖和覆盆式青石柱础，不甘心时光的流逝，向游人证明着这里昔日的兴盛。

The ancient town Zhouzhuang, known as "China's No.1 Water-Town", enjoys its unique geographic conditions of "being surrounded by lakes and streams, commuting by boats" since the old times. Zhouzhuang, with its original name of "Zhenfeng", had been called as "Zhenfengli" since the beginning of Han Dynasty. "Li" was an administrative unit in ancient times, with fifty families in one Li. In the first year of Yuanyou Period of Northern Song Dynasty (A.D. 1086), an official named Zhou donated his house to be a temple. Local people then renamed "Zhenfengli" as Zhouzhuang to honor him. Surrounded by three lakes and one river, the whole Zhouzhuang ancient town is like an island in waters. With only the area of 0.47 square kilometers, Zhouzhuang presents the total waterscape and landscape of "elegantly-crafted bridges, flowing streams and classical dwellings" of the ancient waterfront areas in the south of yangtze River.

During the development of more than 900 years in the past, the ancient town Zhouzhuang has gradually formed the water systems with two longitudinal rivers and two latitudinal rivers. Of the two longitudinal parallel water systems, one is Fu'anqiao River (written as "Nanbeishi River" on Zhouzhuang tourism map and known as "Fu'anqiao River" according to the Zhouzhuang Town Annals of the Qing Dynasty), and the other is Youcheyang River. Among the two, Fu'anqiao River is more important since it passes through the ancient town, connecting the main canal of Jishui Port to the north and the Nanhu Lake to the south. Fu'anqiao River not only acts as the major transport route between the ancient town and the outside, but also introduces the live water from Jishui Port to Nanhu Lake, activating the whole water systems, forming circulating water and sustaining the water quality. Of the two latitudinal water systems, one is called Hougang River, and the other is Zhongshi River. Besides bringing boat-commuting convenience for all residents of the ancient town, these two rivers also supplement vitality to the whole water systems to maintain good water quality.

The origin of Zhouzhuang could be traced back to the Liangzhu Culture Period more than 5,000 years ago. The place of Taishidian nearby the town was once the gathering location of the ancient residents during the Neolithic Period. By making use of the unique geographic waterscapes, the ancient town has produced its own unique living style depending on the advantages of the landscape and waters.

Among the poetic and pictorial water systems, the dwellings and bridges present various styles and rich variations. Every family is by the rivers. Waters and the bridges are also the roads. The traditional buildings such as ancient streets, old lanes, stone bridges, stone wharfs, stone docks, stone railing, bridge-archways, water alleys and water attics are connected harmoniously as a whole so that the crystal-chopsticks-like streams and paths also present the marvelous sight of "sedan chairs entering through the gate and boats passing through home" of the Zhang House.

Among the nearly thousand courtyards in Zhouzhuang, more than sixty percent of the buildings constructed during the Ming Dynasty, Qing Dynasty and ROC period are still preserved. Among them, there are nearly one hundred ancient courtyards and more than sixty brick-sculpture gates and buildings. There are also some street-straddling buildings and water-wall gates that are very typical in the water-towns. Among these buildings, the most representative one should be Shen's House. Shen's House is situated at Nanshi Street that is south to the flat ground at the east end of the Fu'an Bridge. Shen's House faces south, with seven halls. There are more than one hundred rooms with varied sizes; all are distributed on both sides of the one-hundred-meter axis. Shen's House covers the area of more than two thousand square meters and is one of the major cultural heritages of Jiangsu Province. The house was named as Jingye Hall and later changed into Songmao Hall. It was completed by Shen Benren, a descendant of Shen Wansan (the legendary Qing dynasty businessman), in the seventh year of Qianlong Period of the Qing Dynasty (A.D. 1742). Shen's House is composed of three sections. The front section containing the water-wall gate and the river wharf for residents to dock boats and wash clothes is the unique constructions in the water-towns. The middle section includes the wall-gate, tea hall and formal hall, where the hosts meet guests and conduct major events like wedding ceremony or funeral arrangements or issues discussion. The rear section is composed of a big cella building, a small cella building and a back hall room, which are for daily living. The overall house is of the typical architecture layout of "front hall and back cella". The buildings and rooms at the front and back are all connected by street-archways and alley-attics, forming a ring-shape that is very rare among the same type of buildings. The wooden beams of the big cella are of vigorous forms and are all of the round designs from the Ming Dynasty style. The floor boards are firm and solid, and mostly the single-layer pine plates with the width of about sixty centimeters, from which the construction difficulties at that time can be imagined. The railings and windows of the big cella are made delicately, presenting a different architecture style compared with the front hall, belonging to the Anhui style.

Zhang's House, situated on the Beishi Street, is one of the well-preserved residential buildings of the Ming Dynasty in Zhouzhuang. According to the Zhouzhuang Town Annals, Zhang's House was constructed in the Zhengtong Period of the Ming Dynasty by a descendant of Xu Mengqing, a younger brother of Zhongshan King Xu Da. Its original name was Yishun Tang. Xu's family waned at the beginning of the Qing Dynasty and Yishun Tang was purchased by Zhang's family. There are two magnolia trees at front of the gate. In spring, the magnolia trees would blossom and swallows would build nests under the penthouse. Then Zhang's family renamed Yishun Tang as Yuyan Tang, also well-known as Zhang's House. With an area of 1,884 square meters, Zhang's House has five halls and seventy rooms in total. The whole building is of the typical layout of front hall and back cella, with delicate design, fine carvings, ancient and elegant appearance. When entering the gate hall, you could find the fourth hall, also known as the big cella building, where the owner's families were living at that time. The big cella building is of the typical characters of the buildings of the Ming Dynasty, with strong cella columns situated on wooden bases. Even after more than 500 years, they are still intact up to now. The secret rests with the strict operation procedures for the fabrication of the wooden bases. Rough blanks should firstly be carved from solid Nanmu wood, and then they should be immersed into boiling white oil. After absorbing sufficient white oil, the rough blanks should be taken out and further processed into wooden bases. Hence, they are quite lasting. The pair of scrolls of "palankeens entering through the gate and boats passing through the home" in the cella building written by Wu Guanzhong does deserve pondering. East to the cella building, there is a dim and long “assistant lane” . In old times, the front gate would not be opened if there were no major events. Only when there were major events such as wedding ceremony or funerals, or coming guests, would the front gate be opened and then the palankeens could come in. During daily life, the family members would come and go through the assistant lane. Now, the assistant lane has become a tourism channel with a special style. Walking into the dim and long lane, you would see a somewhat green light coming from afar and at the end, you would see a marvelous sight: luxuriant trees and flowers in the backyard, and a crystal river flashing at the terminal of the lane, flowing in alongside the wall and flowing out through the water-attics.

Besides the Zhang's House, there is also the Ze's House which is also a building of the Ming Dynasty. Ze's House is north to the Fu'an Bridge and was constructed by a notable family of Ze Wenhuan during the Chongzhen Period of the Ming Dynasty. Ze's House has five halls in total but there are only the first, fourth and fifth halls left, and the other two have vanished. The main hall is also known as the Big Study Room. According to the records of the "Genealogy of Ze's Family", during the Guangxu Period of the Qing Dynasty, Lu Runxiang began his study here and reached Zhuangyuan, champion in imperial examinations, after growing up. Then the hall where he studied was named as Big Study Room. Now the Big Study Room has disappeared, with only remaining bricks and stone bases lying on the broad homestead, conveying its vitalities in the past to tourists.

II

JINXI 锦溪

有苏南明珠之称的锦溪镇则另有一番特色，锦溪曾因作为宋孝宗为其宠妃陈氏水葬之地而改名为陈墓达800年之久，直至1993年才复名为锦溪，它位于昆山市西南，东与上海市青浦区接壤，西与苏州市吴中区为邻，全镇占地面积90.69km^2，镇区面积1.6km^2。

锦溪被五湖三荡环抱，湖水波光粼粼，因一条美似流霞的小河“锦溪”贯穿全镇、支流如网而得名，锦溪湖荡密布，古桥联袂。五保湖中的陈妃水冢犹似龙口的明珠，龙体有四足，分别为两岸东西各两条河流。古镇布局亦以河道为骨架，街道依水而建，房屋临水而筑，水、路、桥、民居、街市巧妙而自然地融为一体，长达6km的沿河石驳岸、石阶河滩错落有致，逶婉曲折，形态迥异的石驳、河埠、船码头，以及散落着的数以千计造型各异的系缆石，自然地形成了一道十分靓丽的风景线。明代诗人沈周、高启、文徵明等人都为锦溪留下了睹景思情的千古诗文，明代文人高启有诗赞其曰“红杏碧桃花烂漫，长堤曲巷水流漓”，清代的张尚垣也留下“不知谁唱春风曲，顿使桃源绿浪生”的诗句赞咏锦溪。

此外，锦溪的水乡特色也博得了现代著名文人沈从文的青睐，在唐山大地震后的1976年8月间，沈从文先生夫妇带着孙女沈红来到了锦溪度过了一段安静的日子。他的妻子是锦溪人，亲家住在镇北天水桥堍。沈从文对水有着特殊的感情。他说：“我所写的故事，多数都是水边的故事。故事中所最满意的文章，常用船上岸边作为背景。”锦溪的水，与他熟悉的湘西的沅水风格迥异，这里又远离京畿，使他感到无比的安谧。在锦溪的日子里，他常常站在五保湖畔的十眼桥上凭栏而望，静静地看着天空的鸥鸟，湖中渐远渐近的帆影，若有所悟。他觉得湘西的沅水是一个赤膊拉纤的纤夫，锦溪的湖泊就好像睡梦中的少女，淳朴、宁静、清雅，而又含蓄。

锦溪古镇传统建筑以明、清、民国初期为主，具典型的江南建筑风格，古建筑群面积占镇区民宅86%以上，建筑特色古朴，是十分宝贵的文化遗产。古镇以民居四合院为主，以及水墙门、吊脚楼、落水廊棚、桥楼廊坊等独特的河街集市建筑，其房屋多为一两层的砖木结构，青瓦屋顶，为了适应温和而湿润的气候，民居一般为天井、后院、穿堂、陪弄，成院落式、堂式的布局结构，房屋大小、高低不同，具有江南典型的小桥流水风格。

上图：宋孝宗为其宠妃陈氏水葬之地——五保湖，锦溪得名陈墓由此而来。

下图及对页：锦溪古镇以民居四合院为主，以及水墙门、吊脚楼、落水廊棚、桥楼廊坊等独特的河街集市建筑。

Above: The emperor Xiao in Song Dynasty buried his concubine Chen in Wubao Lake, so Jinxi is also named after Chenmu.

Below and opposite page: The old town gives priority to residential Siheyuan, besides water-wall gates, the hanging attics, roofed street, and forebridge etc. particular riverside street market architecture.

Jinxi Town, known as the pearl of south Jiangsu, is of another set of features. Jinxi was renamed as Chenmu for a period as long as 800 years since Emperor Xiaozong of the Song Dynasty had a water-funeral for his beloved concubine Chen at this place and the name was changed back to Jinxi in 1993. Jinxi is southwest to Kunshan City, neighboring Qingpu District of Shanghai on the east and bordering Wuzhong District of Suzhou on the west. The whole area of Jinxi is 90.69 square kilometers, and the area of Jinxi Town is 1.6 square kilometers.

Jinxi is ringed by five lakes and three ponds. It is named after the beautiful "Jinxi Stream" which is passing through the whole area of the town with multiple branches. Jinxi has a dense distribution of lakes and ponds, and also many ancient bridges. The water-tomb in the Wubao Lake seems like a pearl in the mouth of a dragon, with the two rivers respectively to the east and west as its four legs. The layout of the ancient town is also based on the channels of the rivers. Streets and houses are built alongside the rivers. The rivers, roads, bridges, houses and markets on the streets are integrated with each other in a delicate way. The stone embankment, stone stairs, barges, wharfs and docking stones scattered alongside the river are naturally forming a pretty nice sight. Poets of the Ming Dynasty such as Shen Zhou, Gao Qi and Wen Zhiming etc all composed touching poetries at the sceneries of Jinxi. Appreciating the sight at Jinxi, the poet of the Ming Dynasty Gao Qi composed "red apricots, green peaches and blossoming flowers, long embankment, curved lanes and crystal rivers". Admiring the scenery of Jinxi, the poet of the Qing Dynasty Zhang Shangyuan also wrote down "hearing songs of the spring, seeing green waves of peaches".

Additionally, modern scholar Shen Congwen also left his prints here. During August in 1976 after the big earthquake in Tangshan, Shen Congwen spent a quiet time period in Jinxi together with his wife and granddaughter Shen Hong. His wife's hometown is in Jinxi and his parents-in-law lived at the end of the Tianshui Bridge in north of the town. Shen Congwen had a special feeling for water. He once said "stories I write mostly happen nearby the waters and the passages I find most satisfactory in the stories are usually of the boats or embankment as the settings." The waters in Jinxi are of a sharply different style compared with the Ruan River in west Hunan that he is familiar with. Jinxi is far from the capital city, making him feel unprecedentedly peaceful. During his days in Jinxi, he usually stands on the Shiyan Bridge beside the Wubao Lake, looking into the far, quietly looking at flying birds in the sky, the gradually moving boats and seemingly thinking of something. In his opinion, the Ruan River in west Hunan feels like a barebacked boat-tracker, while the lake in Jinxi is just like a sleeping girl, pure, tranquil, elegant and implicit.

Jinxi also has the ancient legend of "thirty-six bridges and seventy-two kilns". There are now twenty-six ancient bridges in the town and they are mostly the buildings of the Ming and Qing Dynasty. The scripts, scrolls and carvings are delicate, presenting the unique "bridge culture" of the water-town. Since the Ming Dynasty, the ancient town had been well-known for its kilns, which consisted the traditional pillar industry for a long period and produced the unique "brick-tile culture". Currently, there are about fifteen ancient kilns of various types in the town. Jinxi has also the only one well-preserved ancient kiln cluster in east China. Jinxi Ancient Brick-Tile Museum, which has a collection of more than one thousand treasures, is the first one of the type in China.

Traditional buildings in the ancient town of Jinxi were mainly constructed in the Ming Dynasty, Qing Dynasty and the primary period of ROC, with the typical architectural style of south regions of the Yangtza River. The ancient buildings account for more than 86% of the residential houses. Buildings are simple, and they are very valuable cultural heritages. In ancient town, buildings are mostly courtyard dwellings, together with unique buildings such as water-wall gates, the hanging attics, roofed street, and forebridge etc. Houses are mostly of the brick-wood structure in one story or two stories, with black-green tiles on the roof. In order to adapt to the warm and wet climate, residents usually have small yards, back yards, open halls or assistant lanes that are arranged in the courtyard pattern or the hall-cella pattern. The houses are of different sizes and heights, and are of the typical style of small bridges and flowing streams.

III QIANDENG 千灯

同样隶属昆山的千灯，也是江苏省历史文化名镇，距今已有二千五百年的历史，古镇物华天宝，人文荟萃，素有"金千灯"之美称。千灯古镇的石板街，南北贯穿古镇，并连接各支路，呈蜈蚣形。全长1.5km，主干街道长800m，由2072块长条形花岗岩铺设而成。石板下面设有既宽又深的下水道，与古镇各处的河埠、河滩相连通。因此，即使下滂沱大雨，石板街上从不积水，转晴即干，让古建筑专家也惊叹不已。

千灯古镇由大户名宅构成的古建筑群中，最值得称道的，是爱国学者顾炎武的故居——顾氏南宅"贻安堂"。南宅现存五进明清建筑，雕梁画栋，砖刻门楼十分精细，具有极高的艺术价值。

千灯古镇另一座保存较好的建筑群是余氏的家宅。余氏从万历年间迁居至此，以经商致富，余宅的建筑特色为前典当、后住宅。现保存双排四进。其中"立三堂"是前后三进六座楼房，相互穿通，呈"亚"字形走马楼，为典型的徽派建筑艺术。

而著名的秦峰塔建于南朝梁天监二年(503年)，系砖木结构，原分7级，高38.7m。现在，秦峰塔俨然已经成为千灯镇标志，还未进镇，远远地映入眼帘的就是直指云天的古塔。因塔身修长，故又有"美人塔"之誉。在秦峰塔下的玉佛殿中，横卧着一尊来自缅甸的号称世界第一大的玉卧佛。这尊玉卧佛长8.9m，高2.45m，重约有30t，镶嵌有1500多颗红宝石、蓝宝石和翡翠，莲花座和衣纹用纯金镏金而成，整座佛像显得华贵端庄，具有很高的艺术价值。这座玉卧佛由于体积空前巨大，已经被载入了世界吉尼斯纪录大全。

千灯古镇的古河埠约有二百多个。它的构筑真可谓千姿百态。镇市中心有宽大的公有码头，专供大船装卸货物。

除此以外，还有特色河埠和大量的独家河滩。有的缩在石驳岸内呈凹式河滩，有的设在楼下，在石框内装木栅门或水榭门封闭。这些河埠，与古镇的贸易和居民的生活关系密切，是千年古镇繁荣昌盛的见证。

宁静而内蕴丰富的古镇，如今正日益焕发出新的活力。昆山文化艺术中心在深刻了解昆山所辖这些江南水乡中独特格局的古镇之经典性、完整性、文化性的基础上，恰如其分地加入了现代因素，整个建筑群沿水体形成自然的进退关系，并巧妙地用一系列自由伸展的曲线形廊桥把周围的体育场馆联为一体，让人们在现代建筑中感受到昆山无处不在的水乡特色。

对页上图：千灯恒升桥河埠。
对页下图：千灯延福禅寺内的秦峰塔为砖木结构，由于塔身苗条，风姿绰约，亭亭玉立，故自古以来秦峰古塔就有“美人塔”之誉，为江南所罕见。
本页上图：千灯浦的两岸，一边是高低错落的江南民居，一边是林荫掩映下的供人休息的廊棚。
左图：千灯古镇的石板街。
下图：千灯的仿古戏台。
Opposite page, above: The old pier in Hengsheng Bridge of Qiandeng town.
Opposite page, below: The Qingfeng Tower in Qiandeng Yanfu Buddhist Temple is made of brick-wood structure; The warped overhanging eaves look like the flapping wings. Thanks to the slim and graceful body of tower, it was honored with the name "Beauty Tower" instead of Qingfeng and rare in region south of the Yangtze River.
Above: On the both sides of Qiandeng River, one is jagged residence of south of the Yangtze River, the other one is roofed street in the shadow of trees which be used to having a rest.
Left: The stone slab street in Qiandeng town.
Below: Archaized opera stage in Qiandeng town.

Qiandeng is also in Kunshan and is another well-known historical and cultural town in Jiangsu Province. With a history of about two thousand and five hundred years, the ancient town of Qiandeng is a land of attractive resources and a birth place of creative minds and gifted talents, and is endowed with the crown of "Golden Qiandeng". The stone-plate street, in the shape of a centipede, is winding through the ancient town, connecting various branch roads. The total length of the stone-plate street is 1.5 kilometers, and the trunk street is 800 meters long, composed of 2,072 pieces of strip-shaped granite plates. Under the stone plates, wide and deep water channels are built to connect with various river wharfs and banks across the ancient town. Therefore even with heavy rains, there would be no accumulated water on the stone-plate street and the street would become dry immediately after it turns sunny, which is also amazing for ancient-architecture experts.

Among the ancient building clusters of notable families and famous figures in the ancient town of Qiandeng, the most well-known should be the "Yi'an Tang", Gu's South House, the former residence of the patriotic scholar Gu Yanwu. The South House has five halls with the style of the Ming and Qing Dynasty. Carvings on the beams, columns, bricks and gates are very delicate, with significantly high artistic value.

Another well-preserved building cluster in the ancient town of Qiandeng is Yu's House. Yu's family moved here in the Wanli Period of the Ming Dynasty and become rich by doing business. The architecture feature with Yu's House rests with its layout of front pawn-shop and back residence. Now there are two lines of rooms and four halls. Among them, "Lisan Tang" is a building of three halls and six cellas that are inter-connected and of the Anhui architecture style.

The famous Qinfeng Tower was constructed in the second year of Tianjian Period of the Liang State in Nanchao Dynasty (A.D. 503). The tower is of the brick-wood structure, with seven stories and the height of 38.7 meters. Now, Qinfeng Tower has become the symbol of Qiandeng Town which you can see from afar outside the town. The tower is of a slim body and thus is given the name of "Beauty Tower". In the Jade Buddha Hall of Qinfeng Tower, there is a reclining jade Buddha that came from Burma and is the largest one in the world. The jade Buddha is 8.9 meters long and 2.45 meters high, with the weight of about 30 tons and embedded with more than 1,500 pieces of rubies, sapphires and jades. The lotus blossom, robes and ornaments are made of gold, making the whole Buddha noble and dignified and be significant artistic value. The jade Buddha has been enlisted into the Guinness World Records for its extensively huge volume.

There are more than two hundred ancient river wharfs in Qiandeng Town and the structures of the ancient river wharfs are of various shapes. At the center of the town, there is the wide public wharf special for loading and unloading goods to and from big boats. Additionally, there are also some river wharfs with unique characters and lots of river embankments exclusive for certain families. Some of them are embedded into a concave river embankment. Some are under the buildings and could be closed with a paling gate or a waterside gate in a stone frame. These ancient river wharfs are of close relation with the trade of the town and the living of the residents, verifying the prosperity of the ancient town in the long history.

The tranquil and profound ancient towns have now been presenting increasing vitalities, and have been integrated together with the modern living of the residents. Based on in-depth comprehension of the classical, complete and cultural characters of the unique ancient towns, Kunshan Culture and Art Center properly introduces the modern elements to form a harmonious relation between the whole building clusters alongside the rivers and the nature, and skillfully applies curved archways to integrate the surrounding sports stadiums and forest parks, and further make people feel the unique water-town features from the modern buildings across Kunshan.

文化传承 Cultural heritage

文化艺术与民间技艺的传承

昆山文化艺术中心在渊源广博的文化背景下，植根于这片沃土中，为这个城市和居住在这里的市民提供一个更适合传承文化、发扬传统的舞台

昆山历史悠久，物产丰盛，环境优美，人杰地灵，众多因素使这处灵秀的江南古城成为众多传统文化艺术与手工技艺发展繁衍的舞台。众多文化与艺术元素的积淀蕴生出内涵丰富的昆山精神特质，为昆山人留下了宝贵的精神文化遗产。

INHERITANCE BETWEEN CULTURAL ART AND FOLK SKILLS

UNDER THE DEEP AND EXTENSIVE CULTURAL CONTEXT, THE PROJECT HAS DEVELOPED TO OFFER A STAGE THAT IS MORE SUITABLE TO INHERIT AND PROMOTE CULTURE AND TRADITION TO THE RESIDENTS OF THIS CITY

With long history, rich resources, beautiful environment, the outstanding people and smart land, the delicately beautiful ancient city of region south of the Yangtze River has become a stage of development in numerous traditional culture art and manual skills. The Kunshan essence with rich connotation has been created by numerous culture and art factors to leave the valuable spiritual cultural heritage for people.

I CULTURE 文化

昆山的文化艺术，首屈一指的是昆曲。这里是百戏之祖——昆曲的发源地。2001年5月18日，联合国教科文组织宣布第一批“人类口头和非物质遗产代表作”名单，中国昆曲艺术名列其间。中国成为首次获此殊荣的19个国家之一。经由唐代黄幡绰、元末顾坚和顾阿瑛、明代魏良辅、梁辰鱼等剧作家、作曲家的不懈努力，昆曲在昆山这片沃土完成了昆山腔——昆曲——昆剧的演变过程，日臻完善，攀上了戏曲艺术的峰巅。许多地方剧种，包括京剧、晋剧、川剧、赣剧、桂剧、邕剧、越剧和粤剧等等，无不受到昆曲艺术多方面的哺育和滋养。

昆曲在原昆山腔的基础上兼收南北曲之长，“其排腔、配拍、榷字、厘音，皆属上乘”。无比委婉动听的乐曲，巧妙地体现了东方人的情感特征，许多难以表述的情感，水一般地在委婉悠长、顿挫抑扬中汩汩流注，打动着人们内心世界最柔美的感区，接收者的陶醉、融化、沉缅、震颤，已非语言可形容。于是天下翕然宗之。

依字行腔的昆曲唱腔，十分讲究语言。语言中包含的字音、声调、语调是构成旋律节奏的基础。所以唱曲首先要识字。为了唱准字音，昆曲不使唱倒字。在曲调的行腔上有四声腔格规范，在度曲时有特定的口法，比如嚯腔、豁腔、断腔等。为了保证字正腔圆，还引申出许多常用的装饰腔格式如摟腔、橄榄腔等。

昆曲的唱腔艺术，以角色行当（生、旦、净、末、丑）分类，这与西方音乐以音域分类有很大的区别。不同行当真假嗓音的运用、共鸣位置的处理，也有独到之处。这样做，形成了各个行当典型的音色，对于塑造人物形象十分有利。净角显得雄浑厚亮，老生显得宽厚苍劲，而小生则用真假声相结合的龙凤音，显得刚柔相济，旦角却是娇润脆亮，莺声呖呖。自然，根据角色个性的需要，演员也要调整自己的音色，以不同的技巧塑造不同的形象。

昆山腔开始其流布区域，只限于苏州一带，明代万历年间，以苏州为中心扩展到长江以南和钱塘江以北各地，并逐渐拓展到福建、江西、广东、湖北、湖南、四川、河南、河北各地，万历末年还传入了北京。到了清代，由于康熙喜爱昆曲，更使之流行。昆山腔成为从明代中叶至清代中叶影响最大的声腔剧种。

不少专家认为，“昆曲所代表的美学趣味虽然明显是南方的，尤其是江南地区的，但是其文化身份却并不属于一时一地，它凝聚了中国广大地区文人的美学追求以及艺术创造。正是由于它是文人雅趣的典范，才具有极强的覆盖能力，有得到广泛传播的可能，并且在传播过程中，基本保持着它在美学上的内在的一致性。”

作为推动昆曲重新迸发活力的重要舞台，文化艺术中心在设计中，巧妙汲取了昆剧表演中的甩袖形象，衍生而成了建筑的整体平面形态，建筑从一个中心逐渐旋转发散，如昆曲表演中翩翩舞动的长袖。通过类似对这种神韵的把握运用，使文化艺术中心这个现代建筑和昆曲等昆山传统文化艺术有机地联接在了一起。

此外，宣卷也是一种独特的民间曲艺形式，以古镇锦溪最具盛名。它兼有说唱、评弹之风，分丝弦宣卷与木鱼宣卷两种。丝弦宣卷一般由六人组成宣卷班子，操二胡、三弦、笛子、木鱼、铜磬等乐器。演出时以传统昆曲唱腔、民间小曲《四季调》掺合申曲、锡剧等地方戏调，悠扬动听，抑扬顿挫。宣卷主角长衫广袖，手执折扇、惊堂。木鱼宣卷则常由两人搭档。一人敲着木鱼，边唱边白；一人和击佛磬，口念佛号以和卷，形式相对简单。

Kunqu opera is the most popular cultural art in Kunshan. It boasts the grandfather of many dramas-the birth land of Kunqu opera. On May 18, 2001, Kunqu opera was selected in the list of the first "Masterpieces of the Oral and Intangible Heritage of Humanity" declared by United Nations Educational, Scientific, and Cultural Organization (UNESCO). China has become one of the 19 countries that won this honor. Through unremitting endeavor by Huang Fanchao of Tang Dynasty, Gu Jian and Gu Eying in the Later Yuan Dynasty, Wei Liangfu and Liang Zhenyu of Ming Dynasty, Kunqu opera has developed from Kunshanqiang and Kunqu opera to Kundrama; it has been increasingly improved to reach its peak. There are many local operas, such as Beijing opera, Iin opera, Sichuan opera, Jiangxi opera, Guangxi opera, Shaoxing opera and Guangdong opera, which have been cultivated and enlightened by Kunqu opera. Nowadays, performers of Kunshan have played a fundamental role in Kunqu opera theaters in China.

Based on the original Kunshan Qiang, Kunqu has integrated strong points both of southern and northern drama. “Its series singing, matching beats, polishing words, determining tone are excellent”. Very pleasant euphemism music subtly reflects the emotional characteristics of Easterners. Many unspeakable emotions run gurgling as water in the circumbendibus and long euphemism with cadence patterns, touching the gentlest sensory area of people's inner world, the recipient intoxicated. Euphoria, indulge, tremor, and wallowing cannot be described by words. Thus it has become very popular over the world.

In line with spelling, Kunqu singing pays particular attention to the words. Pronunciation, tone and intonation contained in words constitute the basis of melodic rhythm. Therefore, it is required to learn to read before sing it. There are four tune patterns in the rendering of an operatic tune. During singing according to the music score, there shall be specific mouth singing, such as close singing, open singing and snapped singing and so on. To ensure enunciation, it also forms many popular beautified formats such as chorus singing, olives singing and so on.

Art of singing of Kunqu is classified by role types (Sheng, Dan, Jing, Mo and Chou), which are very different from Western opera. The resonance processing, position of real tone and falsetto used by different roles are also unique. By doing so, it forms the typical timbre of each role and is very favorable for portraying characters. Jing sings deep and resonant; Laosheng sings old and strong; Xiaosheng sings hard and soft by combination of real tone and falsetto; Dan sings clear and sharp just like warbling of the oriole. As a matter of course, the actors have to adjust their voices in line with personalities of roles, using different techniques to shape different characters.

At the beginning, Kunshanqiang boomed only around Suzhou. In Wanli Era of Ming Dynasty, it spread the areas across in the region south of the Yangtse River and in the north of Qiantang River, and thereafter extended to Fujian, Jiangxi, Guangdong, Hubei, Hunan, Sichuan, Henan, Hebei etc.. At the end of Wanli Period, it was also introduced to Beijing. By the Qing Dynasty, as Emperor Kangxi was fond of opera, then it become more popular. Kunshanqiang becomes the most influential vocal cavity opera from the middle period of Ming Dynasty to the middle period of Qing Dynasty.

Many experts believe that "Although Kunqu represents the Southern (especially in south of the Yangtse River) aesthetic taste, but its cultural identity does not just belong to one place or one era, but it embodies the aesthetic pursuit and artistic creation of Chinese literati from vast areas. As it is a model of literary elegance, it has strong coverage and possibility to be widespread, and in the communication process, it basically maintains the intrinsic aesthetic consistency."

Xuanjuan is a unique folk art form popular in Jiangsu, Zhejiang and Shanghai, most prestigious in ancient Jinxi town. It combines Chinese musical dialogue, storytelling and ballad singing. By the numbers of singers and complication and simplicity of performing, it is divided into string Xianjuan and wooden fish Xuanjuan. String Xianjuan team commonly consists of 6 members, performing erhu, sanxian, flute, Wooden fish, copper chime and other musical instruments. During performances, based on the traditional Kunqu singing and folk ditty "Four Seasons tune", combined with other local opera tune such as Shanghai opera, Wuxi opera and so on, it sounds melodious and cadence. The leading role of Xuanjuan wears wide sleeves gown, with a folding fan and a gavel in hand.

开篇：中国锦溪宣卷艺术馆内具有江南园林特色的庭院景观。**对页上图：**金石人家篆刻艺术馆内部景观。

对页下图：锦溪柿园内景，它是近代著名画家、围棋国手陆曙轮先生故居，因园内有两株柿树而得名。现为已故老人陆曙轮先生和他的次子陆家衡先生的书画艺术陈列馆。

本页：中国收藏艺术展览馆所展示的藏品多为明清紫檀木、花梨木制作的名贵家具。小小的展馆洋溢着一股浓郁而沉重的古典氛围。

Opening page: China Jinxi Xuanjuan Gallery enjoys the characteristics of garden landscape on the Yangtze delta.

Opposite page, above: Interior view of Jinshi Family Sealcutting Art Gallery.

Opposite page, below: Jinxi Persimmon Garden is the former residence of recent famous painter and Weiqi player Mr. Lu Shulun, which was well-known as two persimmon trees. It is the calligraphy and painting display hall of deceased Mr. Lu Shulun and his second son Mr. Lu Jiaheng.

This page: Many valuable furniture pieces showed at China Artistic Collection Exhibition Hall are made of redsanders and rosewood during Ming and Qing Dynasty. This small exhibition hall is permeated with the rich and classical atmosphere.

II

HANDCRAFT ART
手工技艺

《天工开物》中有关于制砖的描述："砖：凡埏泥造砖，亦掘地验辩土色，皆以黏而不散、粉而不沙者为上。汲水滋土，人逐数牛错趾踏成稠泥。然后填满木框之中，铁线弓戛平其面，而成坯形……凡砖成坯之后，装入窑中。凡烧砖有柴薪窑，有煤炭窑……" 有着"36座桥，72只窑"美丽传说的锦溪，其制砖烧窑的历史滥觞于春秋，鼎盛于明清。这样的制砖工艺，延续了几千年。至今，在古镇锦溪仍保留着传统的制砖方式与土窑。成为昆山诸多民间手工艺的代表。在没有发明铜与铁之前，取之不尽用之不竭的泥土，是最受欢迎的。从泥土到陶器到砖瓦，这是一条文明之路。聪慧的先人，依靠砖瓦解决了生活中的很多难题。供奉神明的庙宇，抵御外侮的城墙，让灵魂安息的墓室，无一不用砖瓦砌成。锦溪由此建成的"中国古砖瓦博物馆"也早在上个世纪的八十年代亦蜚声国内外。博物馆前身"大东砖瓦厂古砖瓦陈列馆"创建于1980年10月，曾被舆论媒体称之为"国内一绝"、"中国之最"，同时也赢得了国内外专家学者的普遍关注。如今，砖瓦制作技艺已被列为昆山市级非物质遗产名录。锦溪人将中国古砖瓦博物馆内的珍品复制后，作为旅游纪念品。

昆山独有的昆石为世人所瞩目，它又名昆山石。产自玉峰山中，玉峰山仅高84m，所以产量极其有限。大约在一亿八千万年前，由于受到地壳运动的挤压，使地下深处岩浆中富含二氧化硅的热熔液侵入岩石裂缝，冷却后便形成石英矿脉，昆石正是在这石英矿脉晶洞中生成的石英结晶块，主要成份有二氧化硅99.46%，三氧化二铁0.44%,氧化钠0.08%，氧化钙0.02%，莫氏硬度七度。它的石体很硬，硬度近似玛瑙，却又很脆，很难着手加工，所以我们见到的昆石，大多数为独石，在放大镜下能见到密密麻麻的小晶体排列。它的历史比喜马拉雅山还要早，按照石英晶簇、脉片形象结构特征，人们将它们分成鸡骨、胡桃，雪花、杨梅、荔枝、海蜇等十多个品种。上品昆石遍体雪白晶莹，窍孔遍布，玲珑剔透，坚硬如玉，独具"伛"、"醉"、"透"之形态。昆石的加工大致要经过选坯、曝晒、冲汰、雕琢、浸泡等复杂工艺，方能完成。为了加工一块完美的昆石，不惜花费一年甚至更长的时间，不知多少遍地用碱水、海棠花汁和草酸浸泡洗涤，才使它冰清玉洁，玲珑剔透。明清时期，昆石便与太湖石、灵璧石、英石并称为"四大名石"。昆山亭林园内有两座一人高的昆石立峰，为明代旧物，一为"春云出岫"，一为"秋水横波"，窈窕玲珑，窍孔遍布，具瘦漏透皱之态，是仅存的巨峰佳品。昆石精品近年来在国内外举办的奇石展览中展出，多次获得较高的奖项。

在昆山的水乡古镇常常可以看到夏布。那是一种门幅较狭的手工织就的布料，由于麻棉混纺，含有较多的麻，又用淀粉浆过，有很好的透气性，适宜在夏天穿着。水乡潮湿闷热，夜晚多蚊子飞虫，人们用夏布制成帐子，便可以安然入梦。制成衣衫，也让人感到分外凉爽。这种织布是很讲究技术的，一般由中老年男人承担。在织布前，他们先把麻缭绕在杼子上，以便织布时放入梭子。同时在经布机上系好经线，刷上薄浆。木制的织布机有四只脚，犹如前矮后高的木床。前端有一块木板，这就是织布人坐的地方。织布人左手推布筘，右手掷梭子，两只脚踩着踏板，起早摸黑地劳作。织成一匹夏布，往往需要五六天时间。

除了夏布，还有土布。清代陶煦的《周庄镇志》记载："缫布，以棉纱为经，以白缫为纬，而织成者。基子布，白棉纱间以青棉纱织作小方块，而成基盘纹，可为手巾。雪里青布，以青白棉纱逐一相间织成者，又有名芦菲戗、柳条青者，皆青白相间成纹。"这些土布是人们手工织造的，价廉物美，也符合大众的使用习惯。如今，在古镇街头仍能买到土布制成的服装，让人获得返璞归真的体验。

今天在锦溪、千灯等古镇依然有众多体现这类文化传承的博物馆、纪念馆，包括宣卷艺术馆、古砖瓦博物馆、东俊根雕馆、近现代民间壶具馆、宜兴紫砂博物馆、古董馆、奇石馆、明清家具馆，顾坚纪念馆，柿园等，有十几家之多。它们承载着人们对传统技艺的精华的美好回忆，同时，昆山市在现代社会的发展中也越来越需要更为丰富全面的文化载体，昆山文化艺术中心也正是在这种背景下，植根于这片具有深厚文化底蕴的沃土中，以多样化的功能形式，以全新的面貌为这个城市和居住在这里的新老昆山市民提供一个更适合传承文化、发扬传统的舞台。源自昆山的昆曲更将借此契机，重回到昆山这座全新的城市，焕发出她历久弥新的艺术光彩与活力。

—

文：陈益 昆山市文化发展研究中心学术总监，中国作家协会会员

Quoted from *Heavenly Creations* about the description on brick making: "Brick: to make brick by mixing water with clay mud, it also is required to dig earth and test soil samples. Good earth shall be sticky together, and in the form of powder instead of sand. Apply water to wet the soil, drive couples of cattle to toe and tread into thick mud. Then fill the thick mud into a wooden frame, level its surface with a wire bow to form adobe... after adobe being shaped, load into a kiln. For brick making, there are firewood kilns, coal kiln...". With the legend of "36 bridges, 72 kilns", the brick making and kiln building history of Jinxi originated from Spring and Autumn, at its best in the Ming and Qing dynasties. Such brick process has been lasted for thousands of years. Until now, the ancient Jinxi town still has been retaining the traditional brick making methods and earthen kiln, which becomes representatives of many arts and crafts of Kunshan. In the absence of invention of copper and iron, inexhaustible soil is the most popular. From the soil, to pottery, then to the brick and tile, this is a civilized way. Relying on bricks and tiles, our intelligent ancestors have solved many problems in life. Temples dedicated to the gods, city walls for resisting foreign aggression, tombs for souls resting in peace are all made of bricks and tiles. Jinxi has thus built Chinese Ancient Brick & Tile Museum. It is also renowned at home and abroad back in the eighties of the last century. Founded in October 1980, as the Museum's predecessor, "Dadong Ancient Brick Brickworks Museum", has not only been named by public media as "A Unique one in China", "No. 1 in China" etc., but also won the widespread attention of domestic and foreign experts and scholars. Today, brick productive techniques have been identified as Kunshan municipal intangible heritage. Jinxi people have copied treasures Chinese Ancient Brick & Tile Museum as tourist souvenirs.

The Kun Stone is very famous in the world, and it is also named Kunshan Mountain Stone. It originates from Mount Yufeng. Mount Yufeng is only 84 meters high, so its output is very limited. About 180,000,000 years ago, owing to the extrusion by the earth movement, the solution heat deep in the magma underground containing rich silicon dioxide intruded into the fracture of the rock; after cooling, a quartz reef came into being, and Kun Stone is just the quartz crystal ingot generated from this quartz reef. Its main components are silicon dioxide (99.46%), ferric oxide (0.44%), sodium oxide (0.08%), and calcium oxide (0.02%). Its Mohs hardness is 7 degrees. The stone body is very hard, similar to agate, but it is also very fragile. So, it is very hard to process. Thus, most of the Kun Stones we see are monoliths, and under the magnifier, we can see thickly dotted tiny crystals one row after another. Its history is even earlier than Himalaya Mountain. In terms of the structural feature of the drusy quartz and vein formation, people classify them into about a dozen of types, such as chicken bone, Hu pagoda tree, snowflake, waxberry, litchi, jellyfish and so on. The top-grade Kun stone is snow white and crystalline, full of holes, looking very exquisite, as hard as jade, showing various shapes such as "curving", "drunken", and "looking through". The processing of Kun stone is very complicated, including various steps, such as material choosing, insolation, washing, carving, soaking and so on. In order to process a piece of perfect Kun stone, people even spend one year or a longer time. They do immersion cleaning with alkali water, Chinese flowering apple juice and oxalic acid for many times before making it look pure, noble, exquisite and elegant. As early as the Ming and Qing Dynasty, Kun stone was reputed as one of the "Four Famous Stones" together with Taihu stone, Lingbi stone and Ying stone.Within the woodland garden of Kunshan Mountain, there are two Kun stone sculptures as high as a human, left by the Ming Dynasty. One is called "Chun Yun Chu Xiu (Spring Cloud Coming Out of the Cave)" and the other is called "Qiu Shui Heng Bo (Autumn Water in Crossing Waves)". They are very exquisite, full of holes, having the features of "thinness, leaking, penetrating and crimpling", being very rare huge stone works left from the ancient times. Excellent Kun stone works have been shown in many strange stone exhibitions at home and abroad in recent years, and have won high awards for many times.

In the water country and ancient towns of Kunshan Mountain, you can often see "summer cloth". It is a hand-woven cloth with a narrow width. As it is a work of mixed spinning of hemp and cotton, it contains a lot of hemp, moreover, it has gone through the starch slurry process, so, it has good air permeability, suitable for wearing in summer.The water country is damp and sultry, and at night, there are many mosquitoes flying. So, people use this grass linen to make mosquito nets, and they can sleep peacefully. When it is made into clothes, it can also have a very cooling effect. This kind of weaving is very demanding in technique, and usually is done by old people. Before weaving, they reel the hemp around the reed, so that the shuttles can be inserted when weaving. At the same time, they fix the warp threads on the weaving machine, and add the thin paste. The wooden loom has four feet, just like a wooden bed that is low in the front and high in the rear. At the front there is a wooden board, and that is where the weaving worker sits. The weaver pushes the cloth reed with his left hand and throws the shuttles with the right hand. His two feet are treading on the pedal. He will work from early morning till late night every day. The weaving of a bolt of summer cloth usually takes five or six days.

Besides the summer cloth, there is another kind of native cloth called "tubu". In *Zhouzhuang Town Records* written by Tao Xi of the Qing Dynasty, there is such a record: "The native cloth is made of cotton yarn and white cai. Jizi cloth is in the shape of small squares woven with white cotton plus black cotton yarn, and can be used as towel. Xuliqing cloth is made with black and white cotton yarns interwoven together. It is also called 'Lufeiqiang' or 'Liutiaoqing'. It carries stripes of black and white." These kinds of native cloth are hand-woven, with low price but fine quality, so, they meet the common people's needs. Nowadays, in the street of the ancient town, we can still buy clothes made of native cloth, so, we can taste the feeling of returning to nature and simplicity.

Today, in the old towns such as Jinxi and Qiandeng, there are still many museums and memorial halls carrying on these old cultures, such as Xuanjuan Art Gallary, Antique Brick & Tile Museum, Dongjun Root Carving Museum, Museum of Folk Tea Pots of Modern Times, Yixing Red Porcelain Museum, Curio Museum, Strange Stone Museum, Museum of Furniture of Ming and Qing Dynasty, Gu Shu Memorial Hall and so on. The number is as many as more than a dozen.They are bearing people's fine memories of the essence of the traditional arts, and at the same time, Kunshan City is also in dire need for richer and more comprehensive cultural carriers in its process of development of modern society. It is just in such a background that Kunshan Culture and Art Center is rooted in this fertile soil with rich cultural connotations, uses variegated function formalities and brand-new images to provide a more suitable stage for the new and old citizens living in Kunshan City to carry on their cultural traditions. The Kunqu opera stemming from Kunshan will also take this chance to return to Kunshan, this brand-new city, to display its new artistic splendor and vitality that will last forever.

—

Text: Chen Yi, general director of Kunshan Municipal Center of Cultural Development and Research and member of China Writers' Association.

对页，从左至右：据明朝《昆山县志》记载，经典昆曲剧目《牡丹亭》的作者汤显祖也曾出自对顾坚的敬仰来过昆山千灯，图为千灯镇为纪念此事及汤显祖的这个经典剧目而建的牡丹亭。昆山千灯镇的千灯馆收藏了一千五百余盏造型各异的古灯，浓缩了华夏五千年的灯文化，也向人们展示了一部人类的文明史。在昆曲发源地的千灯镇，由著名戏剧表演艺术家梅葆玖先生题写的照壁。

Opposite page, from left to right: According to the *Record of Kunshan County Annals*, Tang Xianzu, the author of classic Kunqu repertory The Peony Pavilion had been to Qiandeng, Kunshan, to show his esteem to Gu Jian. Photo shown here is the Peony Pavilion, which was built to commemorate that play. The Qiandeng Museum in Qiandeng town collects one thousand and five hundred old lanterns with various shapes, which condense the culture of Chinese lantern over a span of 5000 years and exhibits a history of human civilization. The inscription on the wall is written by Mei Baojiu, a famous performing artist, in Qiandeng town as the cradle of Kunqu opera.

曹勇，新华报业传媒集团《昆山日报》副总编辑

Cao Yong, deputy editor-in-chief of Kunshan Daily of Xinhua Daily Press Group

Interview with Cao Yong

City Spirit and Blueprint

城市精神与城市蓝图·曹勇

新印象：在您眼中昆山在适应城市发展和越来越丰富的市民生活要求过程中，发生了哪些变化？

曹勇：昆山因为能够适应改革开放的契机，而主动成为了中国改革开放的十八个典型地区之一，利用了开放的机遇发展了外向型经济。在很长的一段时期内政府提出昆山就是开发区，开发区就是昆山。

在城市发展改善基础设施的过程中，我们也借鉴了南方其他城市的经验教训，如果一座城市只有工业设施没有服务设施，缺乏配套功能设施，那么它就是一座没有生命的城市。以花桥为例，作为软件外包产业基地，很多法国、印度员工随外企进驻，如果没有相应的如健身、与世界院线同步的影城等生活配套设施的规划，工作之余的生活会让他们无所适从，毕竟对他们来说这不是仅仅几个月的时间要打发，没有承载城市生活功能的地方，留不住人才。因而在转型升级时不应该仅仅是经济现代化，仅仅是一个城市规模的转型升级，也应是一个城市功能的转型升级，最终实现人的精神文化的现代化才是真正意义的现代化。所以近年通过昆山城市的总规划，提出“一核五城”的方案。在各个组团中完善城市功能项目，每个片区都在增补一些原来没有的与商业发展，与生活相关的设施。比如东部新区根据发展需要增加了有展览展示功能的国际会展中心，在生活功能方面，杜克大学已经建成，在黄埔城市花园对

New Impression: What changes have Kunshan taken place in your eyes in the process of adapting to urban developments and richer demands?

Cao Yong: As Kunshan can adapt to the opportunity of reform and opening up, it has become one of the 18 typical areas of China's reform and opening up actively, making use of the chance of opening up to develop its export-oriented economy. In a long period of time, the government puts forward that Kunshan is a development zone.

In the process of urban development and infrastructure improvement, we also learn from the experience and lessons of other southern cities. If a city only has industrial facilities while lacking of service facilities and supporting functional facilities, it will be a lifeless city. Taking Huaqiao as an example, as a software outsourcing industry base where many French and Indian staff enter with foreign companies, if there is not the planning of corresponding supporting living facilities, such as fitness facilities and film cinemas which are synchronized with world theater chain, staff will have nothing to do after work, and it is hard for them since they have to stay here for a few months. A city without living functions can not keep talents. Therefore, in transformation and upgrading, it should not only be a matter of economic modernization, but also the transformation and upgrading of urban functions for the transformation and upgrading of urban size in a city. Modernization in its true sense should achieve the modernization of spiritual culture of human beings. Hence, in recent years, the scheme of "one core, five cities was proposed through Kunshan's overall urban planning. In the projects of perfecting urban functions in various clusters, every district is adding some new facilities related to commercial development and life. For example, there is an International Convention and Exhibition Center newly built in Eastern New Area in accordance with development needs. In the aspect of living functions, Duke University has been built, and the Jen Ching Memorial Hospital in Taiwan was introduced in the opposite of Huangpu Urban Garden and the southern side of Qianjin Road. The Sports Center, including a stadium, a gym, an indoor pool, an outdoor football field, etc., was built in the western sub-center. It is planned that a city parlor with gradually perfected functions will appear after the Culture and Art Center has been built and an entertainment complex has been formed and adding the original Forest Park. At present, a commercial complex integrating various kinds of functions and a general hospital are still lacking. Now, the parties concerned are contacting to introduce Show Chwan Hospital, one of the top three hospitals in Taiwan.

面，前进路南侧引进了台湾宗仁卿医院。西部副中心建成了体育中心，包括体育场、体育馆、一个游泳馆、室外足球场等。而现在规划建成了文化艺术中心，形成了文体活动综合区域，加上原有的森林公园，一个城市客厅的功能逐渐比较完善地呈现在大家面前。目前还缺乏一个集合各种功能的商业综合体和一个综合性医院，目前有关方面正在联系引进三甲医院——台湾秀传医院。

新印象：从城市建设的角度，昆山在赢得一系列引以为荣的城市名片之前和过程中，做出了哪些有益的尝试和艰辛的努力，有哪些值得同类城市借鉴的经验?

曹勇：最大的尝试与亮点我认为就是坚持规划在这个城市中的引领作用。昆山这些年一直按照城市的总体规划有计划地进行，一张蓝图一直规划到2030年。而不是像某些大城市在摊大饼一样，城市想扩张到哪里就到哪里，我们参观国内很多地方，都有自己的规划，但是随着领导层的更换却不断地在修改。昆山不存在这种情况。

其次，城市的土地储备制度做得非常好。比如很多地块发展没有成熟，就不急于卖出换取利润。在转型过程中，一批落后的厂房需要淘汰，由政府出资收购回来进入土地储备中心。会根据城市需要理性有序的合理利用，如建学校，建医院，建文体中心，文化生态公园等。

新印象：在为这个城市提供改造和发展的过程中哪些令您印象深刻?

曹勇：最大的感受是我们这座城市所做的功能项目没有贪大求全或标新立异。建筑风格也能很好地融入本土特色。无论是设计成琼花形状的体育场，还是汲取文化元素昆曲水袖的文化艺术中心，以昆山三贤命名的道路及即将兴建的祖冲之公园都源自这种理念。包括道路绿化也没有急功近利，浪费水土资源。选择的都是适合当地环境的树种如香樟、玉兰树、桂花树。这座城市秉承的是一种务实的心态。城市的精神和理念都是一脉相承的。

New Impression: From the perspective of urban construction, what beneficial attempts and arduous efforts have Kunshan made in the process of winning a series of city cards to be proud of, and what experience can be learned by similar cities?

Cao Yong: The biggest attempt and highlight, in my point of view, is to stick to the leading role of the planning in this city. Instead of expanding anywhere at will like some large cities, in these years, Kunshan has developed designedly according to the overall planning of the city which is a blueprint to 2030. Many places in China where we visited have their own planning, but the planning is amended constantly with the changes of leaders. However, Kunshan does not have this kind of cases.

In addition, the city land reserve system is perfect. For example, many plots which have not been well developed will not be sold in a hurry for profits. With the traffic improvement and urban integration, in the process of transformation, a batch of outdated plants needed to be obsolete are bought by the government to enter into the land reserve center, orderly and reasonably according to city needs, for example, building hospital, culture and sports center, cultural and ecological park, etc.

New Impression: What have impressed you in the process of providing transformation and development to the city?

Cao Yong: The greatest feeling is that the functional items done in this city do not strive for grandiose projects or seek to be different. The architectural style can blend in native characteristics well. The stadium designed in the shape of Viburnum macrocephalum, Kunshan Culture and Art Center which extracts cultural elements and actor's long sleeves in Kunqu opera, the roads named after Kunshan Three Worthies, and the Zu Chongzhi Park to be built are originated from this idea. Road greening does not seek for quick success and waste land and water resources. What we chose are trees adapting to local environment, for example, camphor tree, magnolia tree and sweet osmanthus tree. What we persisted is a pragmatic mentality. The spirits and ideas of the city are related.

许振敏，昆山市规划局局长

Xu Zhenmin, general director of Kunshan Urban Planning Bureau

Interview with Xu Zhenmin

City Card and Kunshan Spirit

城市名片与昆山精神·许振敏

New Impression: What changes have Kunshan taken place or are Kunshan taking place in your eyes in the process of adapting to urban developments and richer demands?

Xu Zhenmin: As Kunshan can adapt to the opportunity of reform and opening up, it has applied to become one of the 18 typical areas of China's reform and opening up actively, making use of the chance of opening up to develop its export-oriented economy. Kunshan is not only the first in middle-sized and small cities, but also the top one in Forbes Chinese best county-level cities for 4 continuous years with its strong comprehensive strength. It is also among the best in prefecture-level cities. However, by last year, its service industry had accounted for 39.7% in the industrial structure, being less than 40%, while its industry had had a very large proportion. It is hard to say that a city is a modernize city if its service industry fails to reach 50%. Since its comprehensive strength is strong while the service industry is less-developed, it needs to transform and upgrade. To develop the service industry, the infrastructures of the city are demanding. As a result, it can not satisfy the current situation. For example, Kunshan Culture and Art Center with an investment in infrastructure of 1.45 billion RMB and the Culture Expo Park under construction which reflects the

新印象：在您眼中昆山在适应城市发展和越来越丰富的市民生活要求过程中，已经或正在发生的变化有哪些？

许振敏：昆山因为能够抓住改革开放的契机，利用开放的机遇发展了外向型经济，而成为了中国改革开放的18个典型地区之一。它不仅在中小城市排名第一，还凭借雄厚的综合实力4年蝉联福布斯中国最佳县级城市第一名 在地级市中也名列前茅。但是我们也要清醒地看到昆山发展的不平衡，到去年为止，它的服务业在产业结构里所占比例只有40%，工业比例非常大。一个城市如果服务业达不到50%，就很难说这个城市是现代化城市，所以要转型升级。要发展服务业，就要加快城市功能建设，如基本建设投资十几亿的昆山文化艺术中心，正在实施的反映稻作文明、良渚文化、江南民俗、昆曲艺术等内容的文化博览园，目的都是要创造宜居环境把高端人才引进来，留下来，做到进得来，留得住。

civilization of rice cultivation, Liangzhu culture, Jiangnan folk customs, art of Kunqu opera and other contents all aim to create a livable environment to keep high-end talents.

New Impression: From the perspective of urban construction, what beneficial attempts and arduous efforts have Kunshan made before and in the process of winning a series of city cards to be proud of, and what experience can be learned by similar cities?

Xu Zhenmin: In the objective set forth at that time, it should learn from Singapore for integral development, Taiwan for technological innovation and South Korea for industrial transformation to reach the economic level of Shenzhen at that time in 10 years and that of Singapore in 20 years. The most important thing in planning is that we developed a new editing and revision scheme on Kunshan' s urban master planning in 2008. While in leadership, it has become more and more rational from planning to implementation, and has had long-term goals increasingly. Therefore, the implementation of the planning will last until 2030. This is my main work in 2007, 2008 and 2009. The concept of "one core, five towns" was proposed in the planning. The five towns are five clusters, including Western New Town which is a living area comprising Forest Park, arts center, enterprise center, administration sub-center, etc., Eastern New Area which depends on the development zone, Zhangpu Town in the southern part and Huaqiao District in the northern part.

This editing and revision scheme of planning won the first prize of overall planning of China Association of City Planning in 2009, indicating that it is a scheme which can stand the test of time. Moreover, many cities have implemented their 20-year planning in 5 years, which indicates that their economies develop fast on one hand and the planning is not so forward-looking on the other hand. Therefore, we have done much on determining economic scale when making the planning. By comparing to cities similar to Kunshan in scale and environment as well as Singapore, South Korea and Shenzhen, we have made clear positioning at that time, that is, large city (the positioning of city size), modernization (the positioning of city goal) and being sustainable (the definition of development pattern).

In recent years, we have developed a series of construction projects from the perspective of planning. The urban function reconstruction implemented by us includes many aspects, for example, the ongoing renovation project of being all elevated and high speed in the center. A series of urban open space projects have been constructed, including hospitals and schools. For example, the project of Kangju Xinjiangnan resettlement houses leads Kunshan' s livable communities in fact. Considering architectural color and form and separation of pedestrian and vehicles, the supporting projects for community life include Loujiang Experimental School, kindergarten, and the adjacent Culture and Art Center which integrates with urban functions and links with communities organically. It has become a model livable city.

We also planned to construct 10 projects in Kunshan at that time, including a university, a park, a hospital, a Culture and Art Center, a Sports Center, a commercial complex, etc.

新印象：从城市建设的角度，在昆山赢得一系列引以为荣的城市名片之前和过程中，做出了哪些有益的尝试和艰辛的努力，有哪些值得同类城市借鉴的经验？

许振敏：当时昆山提出的目标是整体发展学新加坡，科技创新学台湾地区，产业转型学韩国。要十年时间打造当时深圳的经济水平，20年时间达到当时新加坡的经济水平。规划方面最重要的一件事，就是2009年完成的新的昆山城市总体规划修编方案。这件事是从2008年至2009年的两年时间里我最主要的工作。规划提出“一核五城”的概念。五城是五个组团包括西部新城、东部新城、北部新城、南部新城、花桥商务区。

这个规划获得了2009中国规划协会优秀规划设计一等奖。我们在做这个规划的时候对经济规模的确定做了大量的工作。通过与昆山规模相当、环境类似的城市相比较，与新加坡、深圳作比较，当时就提出了明确的定位：大城市，现代化，可持续。大城市是对城市规模的定位，现代化则是对城市目标的定位，可持续则是对发展方式的界定。

近年来，在规划指导下围绕增强城市功能的目的，我们展开了一系列建设项目，如目前正进行的中环改造全高架工程、医院、康居新江南动迁房项目等，实际上在引领昆山宜居小区的方向。从建筑的色彩到建筑的形式，到考虑人车分流，社区生活配套项目包括娄江实验学校和幼儿园，临近的文化艺术中心等城市功能相接轨与社区有机的链接互动关系，成为城市宜居的典范。

Currently, a university, a park, a hospital, a culture and art center and a sports center have been built in advance while the commercial complex is under construction.

In this process, our most important change is to let city lead the development of economy rather than let city support the development of economy. At present, many cities still remain in the latter state where urban economy runs with construction projects, or being called as park construction stage.

New Impression: In these city cards, what honors are prided by people most in your opinions? And why?

Xu Zhenmin: In various honors, the UN Habitat Scroll of Honor Award is the highest award, weighing the heaviest. However, instead of this reason, I am proud of it because the theme of hosting and core meaning of winning are to praise Kunshan' s equal attitude towards all young and old residents. It reflects Kunshan's urban spirits of "opening, innovation and excellence". We applied for it for two times, and won the award in the second time finally according to this featured mentality and in an appealing way. It also makes me think of receiving the leaders of Ministry of Housing and Urban-Rural Development of the People's Republic of China (MOHURD) when they visited Kunshan at that time. In the reception center of Forest Park, when seeing that Kunshan residents enjoy the natural air anion bar in the Forest Park in small groups, the leaders of MOHURD exclaimed that "how pleasing the lives of Kunshan people are". Among them, 70-80% is new people in Kunshan.

我们当时提出的建成昆山十项工程：一所大学、一个公园、一所医院、一个文化艺术中心、一个体育中心、一座商业综合体等，目前大学、公园、医院、文化艺术中心、体育中心都已完成，商业综合体正在建设当中。

在这个过程中，我们最重要的改变是让城市引领经济的发展而不是城市支撑经济的发展。

新印象：您认为在这些城市名片中哪些荣誉最值得自豪，为什么？

许振敏：在诸多的荣誉中，联合国人居奖在其中堪称最高奖项，分量也最重，可我不是因为这个原因而感到自豪，而是这个申办主旨和获奖的核心意义是给予昆山这个城市对所有新老居住者平等态度的一种褒奖。并体现了昆山“开放、融合、创新、卓越”的城市精神。这次人居奖我们申报了两次，第二次申报终于按照这个主打思路以一种很具感染力的方式获得了成功。这也使我联想到当时接待住建部领导来昆山参观，在森林公园接待中心，看到周围三三两两都是在森林公园享受这个天然氧吧的昆山市民，住房和城乡建设部领导连连感叹：昆山人生活状态这么惬意。而这里面百分之七八十都是新昆山人。

New Impression: Why is Kunshan Culture and Art Center chosen to near Forest Park?

Xu Zhenmin: There is not suitable space in the old city for an architectural complex at this scale. Western New Area is the living center for Kunshan. Through constructing Kunshan Culture and Art Center in here, the entertainment and sports areas will be contiguous, mitigating parking space problems in the aspect of traffic and making here become a multilevel, multifarious and all-around center for citizen activities.

新印象：当时昆山艺术中心的选址为什么选在了森林公园旁边？

许振敏：因为这种尺度的建筑群在老城区已经没有适合它的空间，西部新区是我们昆山的生活中心区。把昆山文化艺术中心建在这里，就会使文体活动区域连成一片，使这里成为多层次、多样性的、全方位的市民活动中心。

New Impression: What does Kunshan's adaptive capacity to development stem from?

Xu Zhenmin: Kunshan gets today's achievements with its abilities of introduction and absorption. Kunshan is positioned as an international well-known advanced industrial base, a large city close to Shanghai and a modern city in region south of the Yangtze River. From the perspective of humanistic spirit, Kunshan's humanistic spirit of daring to try and innovate is not gained overnight. In the process of opening up, the people of Kunshan, the poorest agricultural county in the big Suzhou area at that time, experienced the advantages of reform. Although it did not being the first to carry out industrial reform pattern in villages and towns, through investment promotion, Kunshan, a city without any industrial base originally, achieved economic take-off. The first land lease, the first enterprise owned by a sole investor and the first joint venture in Jiangsu Province are all in Kunshan. Meanwhile, the new ideas, management thoughts and lifestyles brought by foreign enterprises also promote the integration of Kunshan people's ideas.

新印象：昆山的这种对发展的适应能力您认为源自它的哪些历史传统或者人文精神？

许振敏：昆山历史悠久，人文荟萃。我们的城市定位是成为国际知名的先进产业基地，毗邻上海的大城市，现代化的江南水乡城市。而从人文精神的角度来讲，昆山这种敢于创新的人文精神不是一蹴而就的，吃螃蟹的第一人是昆山人的祖先，在开放的过程中，当初大苏州地区最穷的农业县昆山人体验到了改革的优越性，尽管在乡镇工业改革模式推行的时候没有走在前面，但通过招商引资，昆山这个原无任何工业基础的城市终于实现了经济腾飞。江苏省第一块土地批注，第一家合资企业、第一家独资企业都源自昆山，同时外来企业带来的新思想、新管理观念、新生活方式也对昆山人的思想起到了互相融合的推动作用。

第二章 Chapter 2

文化·传承·思考
Culture·Inheritance·Thinking

本章悉心梳理了昆山市厚重的人文历史、浓郁的地域特色、经典的文化传承等传统脉络，揭示了建筑师创意思维的由来；并通过采访来展现昆山的城市规划蓝图，勾勒出昆山规划与城市设计的思索之旅及其与建筑个体的相互关系。

This chapter carefully sorts out Kunshan's profound culture and history, strong geographical characteristics and classic cultural development and explores the origin of architects' creative ideas; and through interviews unfolds the Kunshan urban planning blueprint and depicts the relationship between the journey of Kunshan planning and urban design and the individual buildings.

采访：许振敏，曹勇 Interview with Xu Zhenmin, Cao Yong

理念·功能·剖析
Idea·Function·Analysis

为实现“房在园中”的构思意图、建筑造型特点及功能要求，甲方、设计、材料、施工各环节密切合作，在制作模型、实验、模拟，反复测试与论证的基础上，最终不仅完整地实现目标，并为业界提供了很多值得借鉴的设计与实践经验。

To achieve the design idea, architectural features and functional requirements of "room in the garden", architects worked closely with clients in terms of design, materials, construction and other aspects. On the basis of modeling, experiments, simulations, repeated testing and demonstration, they finally achieved the whole design objective, providing the industry with a lot of design and practical experience.

采访与文章：崔愷，阿兰·博尼，李跃文，金照彬，范晓玲 Interview & essay: Cui Kai, Alain Bony, Li Yuewen, Jin Zhaobin, Fan Xiaoling

城市·设计·生活
Urbanization·Design·Life

昆山规划以及文化艺术中心的建造，汲取了国外经典案例的规划思想。不仅给居民带来了福祉，并为昆曲等传统文化的再生提供了一个互动舞台，让建筑自然植入城市，选择了一条适合城市与其居住者们的道路。

Kunshan planning and the construction of Kunshan Culture and Art Center draw on the planning ideas of foreign classic cases, providing not only benefits to the residents but also a platform to Kunqu opera and other cultures to revive and interact with people, implanting buildings naturally into the city with a path suitable for both the city and its residents.

采访：周继春，姚伟宏，韩明 Interview with Zhou Jichun, Yao Weihong, Han Ming

2. Idea·Function·Analysis

理念·功能·剖析

“建筑最宝贵的性质是它的多样化和联想到自然界有机生命的生长，这才是真正建筑风格的唯一目标。”芬兰建筑大师阿尔托的这一观点在昆山文化艺术中心的设计中得到了某种印证。

建筑师基于对昆山这座城市的历史背景，文化渊源及基地周边环境等诸多特点的考量，针对之前参与竞标的设计方案存在的或过于集中孤立、没有充分利用场地的资源，或重视了街道界面、却忽视了河道景观带的问题，提出了“房在园中”的浪漫构思，意在打破用地范围的局限，并全力营造一个开放性的建筑，与周边环境相互关联。并采用景观设计手段，形成较为自然的公园氛围，把建筑看作公园的一部分，也避免了一般剧院空间界面的不连续的问题。

为实现文化艺术中心“房在园中”的设计意图，并达到造型的“曲、柔、透”和亲水性等特点，以及从大剧院本身的功能要求出发，从可持续节能到舞台、声学、结构、幕墙的各方设计师与甲方乃至材料选择、施工建造等各环节专业人士密切合作，通过制作大尺度模型，在大量样板实验和模拟，反复测试与论证的基础上，最终得以完整地实现设计目标，并为业界提供了很多值得借鉴的设计与实践经验。

"The most valuable property of building is its diversity and the ability to associate with the natural growth of organic life, which is the unique goal of real architectural style." Finnish architect Aalto's viewpoint has been proved in the design of Kunshan Culture and Art Center. For other bidding designs proposed: some were too centralized and isolated, and failed to make full use of the land; some design considered the street layout but ignored the river landscape. Accordingly, architects lie on Kunshan's historical background, cultural origin and characteristics of the surrounding environment and many other features and proposed the romantic "room in the garden" idea, intended to break the limitations of land use and build a fully open architecture that is linked with the surrounding environment. They also adopt the means of landscape design to form a more natural park atmosphere and consider the building as part of the park, which avoids the space discontinuity of general theaters.

To realize the "room in garden" design idea of Kunshan Culture and Art Center and achieve the "curving, tenderness, transparency" and riverside characteristics, architects set out from the functional requirements of the theatre. Designers and professionals worked closely with clients in aspects of sustainable energy, stage, acoustics, structure, curtain wall design and material selection and construction etc. Through the production of large-scale model, a large number of model experiments and simulations, repeated testing and demonstration, design objective was fully achieved, providing the industry with a lot design and practical experience.

竞标历程

Competition

设计方案1
设计公司:
上海建筑设计（集团）魏敦山建筑创作研究室

Design Proposal 1
Designer:
Shanghai Architectural Design (Group) Wei Dunshan Architectural Creation Workshop

规划背景与概况

昆山文化艺术中心位于前进西路北侧，市体育中心西南侧，西靠规划道路，北面为一号路和体育馆相邻，东侧与体育中心相望。用地面积约为11.2hm²。基地地势基本平整。该中心目的是建成融会议、音乐、影视、演出、展览、商业一体的综合中心。昆山文化艺术中心建成后将与其周边市政建筑共同构筑和谐的文化生态园氛围，成为昆山重要的视觉景观聚焦点，将进一步提升昆山市的文化品位，以及提升昆山城市综合功能。

设计理念与构思

本设计着重于两个主题，分别是“舞动的风琴”和“艺术的殿堂”。两个主题相互渗透，相得益彰。建筑表皮构造吸取了风琴的韵味，使得建筑造型犹如舞动的风琴放在水岸边上，整体环境构筑了音乐海洋的氛围。使建筑本身与整体环境共同构筑了文化生态园的和谐氛围。

艺术中心的设计首先考虑的是与整个环境的呼应，注重生态理论和仿生技术的运用。并通过简洁的造型和使用现代建筑材料膜结构建造出充满韵律的立面，体现了艺术中心的建筑气质，使其成为视觉的景观聚集点之一。

“艺术殿堂”主要体现在内部空间和建筑外观两个方面。流畅的内部共享大厅使整个区域充满活力，通高的大面积膜结构墙和风琴的韵律相似，又充分地将滨河景观引入室内，给人开阔大气的视觉感受。整个立面充分考虑白天，夜间的艺术形象，做到白天建筑造型简洁纯净流畅，晚间灯火辉煌华丽，如艺术殿堂般供公众参与和欣赏。

设计将使昆山文化艺术中心建设成为艺术殿堂和体现21世纪现代化城市风貌的城市标志性建筑。同时注重文化艺术中心与城市环境的关系，艺术中心的建筑造型以及功能布局的可适性。

建筑造型

文化艺术中心馆的建筑造型以大厅空间和屋盖结构形态为基础进行一定的艺术加工，墙体和屋盖结构采用韵律感极强的空间膜结构，富有韵律的波动，上下变化，犹如舞动的风琴奏响美妙动人的音乐。并且文化艺术中心及贵宾楼整体造型简洁，曲线菱形的体量巧妙地化解了与地段周边的建筑相互紧张的关系。

展览中心的屋顶则采用阳光板和膜结构作为采光天窗，膜结构吸取了昆山市花的意向，抽象表达了琼花在建筑形式上的艺术完美统一。使封闭压抑的室内空间引入大自然的阳光与新鲜空气，达到生态和谐的统一。

建筑形象简洁、大气、有力，颇具一种艺术美的雕塑感，体现了文化艺术与美的基本特质。独特的形体，简洁有力的虚实对比，明亮的色调和光洁的质感，使建筑犹如凝结在一颗小小的水珠上，屹立在昆山这青山绿水中。

对页：透视效果图
Opposite page: Perspective view
本页：总平面图
This page: Site plan

Planning overview and background

Kunshan Culture and Art Center is located on the north side of Qianjin West Road and southwest side of Sports Center of the city, bordering on the planned road to the west, No.1 Road and the Stadium to the north and the Sports Center to the east. It covers about 11.2 ha. The site is roughly leveled.

The Center aims to build a comprehensive center integrating conference, music, film/TV, performance, exhibition and commerce. On completion, Kunshan Culture and Art Center will create a harmonious atmosphere of cultural and ecological park with the surrounding public utilities as a key visual view in Kunshan to further improve cultural taste of Kunshan and comprehensive urban function of Kunshan.

Design concept

This design puts priority on "dancing accordion" and "artistic hall" of two themes. The two themes permeate and benefit each other. Construction of building skin draws charm from the organs. It makes the building shaped like dancing organs on waterside. The integral environment creates atmosphere of a music sea. The building proper and the integral environment jointly create harmonious atmosphere of cultural ecological park.

First of all, the design of an art center takes consideration of echoing to the whole environment and pays attention to usage of ecological theory and bionics techniques. Streamlined shape and film structure as a modern building material are adopted to create the elevation full of rhythm, and embody architectural style, making it one of the visual landscape focuses.

The artistic palace is mainly reflected in the internal space and building appearance. The streamlining internal lobby makes the whole area full of energy. Full-length large-area film-structural wall, similar to the rhythm of the accordion, fully introduces riverside view into the room and provides an open and elegant visual experience. The whole elevation takes full consideration of artistic images of day and night. The building will be simple, pure and flowing on the day and illuminating and luxurious at night like artistic palace for participation and enjoyment by the public.

The design will build Kunshan Culture and Art Center into an artistic palace and urban landmark building of modern urban appearance in the 21st century. At same time, priority is put on relationship between the Center and the urban environment as well as the adaptability of building shape and function layout of the Center.

Building form

Hall of the Culture and Art Center is shaped on basis of the reminding of hall space and roof structure shape for certain artistic processing. The walls and roof structure adopts spatial film structure of high rhythm feeling. It is changing vertically like dancing organs to play beautiful and charming music. In addition, Culture and Art Center and the Guest Building are shaped simple. The rhombic size ingeniously corrects the tension with the neighboring buildings.

Roof of the Exhibition Center adopts polycarbonate sheet and film structure as lighting roof. The film structure draws from the image of municipal flower of Kunshan and abstractly shows the artistically perfection and unity of Qionghua Flower in building style. The isolated and depressed internal space has introduced the natural sunshine and fresh air to achieve harmonious and unified ecology.

The building is shaped simple, elegant and vigorous like a statue of artistic beauty to show basic characteristics of cultural art and beauty. The unique shape, simple and vigorous contrast between reality and empty, bright tone and smooth texture make the building as if concentrated in a small water drop and towering among the green.. .mountains and rivers in Kunshan.

设计方案2
设计公司：
上海九城都市建筑设计顾问有限公司

Design Proposal 2
Designer:
Shanghai Nine Town Design Studio For Urban Architecture

背景

昆山文化艺术中心选址于昆山体育中心地块的西南角，场地内的体育场馆已经建成，主体育场体量巨大而且其主轴线与周边道路形成了一定的夹角关系。

文化艺术中心由三组主要的功能构成：即2000座的剧场和多厅影院、会议中心及展览中心。

总图布局

文化中心的总图考虑了以下几方面的因素：其一是在地块中部留出公园空间，以延续从地块北部森林公园一直到地块南部新城核心区公园之间的城市绿化景观轴；其二是削弱建筑体量感、确立以主体育场为中心的场地秩序。为此建筑的三个部分形成“品”字形关系，其中会议和展览靠近外部道路，两者连成一体、形成矩形体量；剧场则凸入地块中部，其形体采用与体育馆接近的椭圆形，延续体育中心建筑群中建筑与景观的关系类型。

“品”字形结构中不同功能区之间则形成公共性强的开放空间，吸引并引导城市人流穿越建筑进入中心公园。文化中心这样的总图布局策略性地消减了建筑的体量感、与周边建筑群形成呼应关系、并且在地块中心形成最大限度的绿地公园。

剖面设计

文化艺术中心的空间组织以开放性、公共性和适应性为原则，希望建筑能更贴近、更灵活地服务于市民。通过剖面楼层层高的精心组织和错层处理，会议及展览中心的主要功能区与辅助功能区实现了上下分层布局，中间最适宜步行到达的区域设定为公共区。公共区域兼顾了使用要求和开放管理的可能性，可以实现局部封闭和开放管理的兼容。一些使用频繁的设施如多媒体中心与电影院则可以单独对外开放。会议及多功能厅的平面考虑自由分隔的可能性，可以举办会议、演出、展览和宴会等多个种类不同规模的活动。建筑群内部多变的楼层及场地标高系统形成了多条体验丰富的公共流线，鼓励市民接近、驻足和穿越建筑。

立面形式

与内在的功能属性相一致，文化艺术中心的会议展览区形态呈现为一个充满悬浮感的金属盒子，立面材料为穿孔金属板与玻璃；而剧场部分则呈现为具有雕塑感的白色曲线形体量，两者之间形成一种戏剧性的并置关系。在这个方案中，文化艺术中心的整体形态虽然没有一个统一的形式逻辑，但它强调的是与地块整体空间关系的整合，这个思路为其后续的方案推进提供了一个特别的视角。

对页：南立面透视效果图
Opposite page: Perspective view of south elevation
本页：鸟瞰图
This page: Bird's eye view

Background

Kunshan Culture and Art Center is located at the southwest corner of the plot where Kunshan Sports Center sits. On the site, the stadium has been completed. Size of the stadium is large. Its main axle forms an angle with the road nearby. Culture and Arts Center is composed of three functions, including 2000-seat theater and multi-hall cinema, conference center and exhibition center.

Master layout

The drawing of the Center takes consideration of the following factors: firstly, to reserve a park space at center of the plot and extend urban landscaping axle from the Forest Park north of the plot to the New City Core Area Park south of the plot; secondly, to weaken the dimension of whole building and establish site order with the main stadium as the center. As a result, the three parts of the building will form a triangle: conference and exhibition area is close to the outer road and integrated as a rectangular size; the theater protrudes into the center of the plot, which is shaped oval similar to the stadium, to follow the relationship between building and landscape in the sports center complex. In the triangular structure, open spaces with high commonness are formed between different function areas to attract and guide urban stream of people to the Central Park through the buildings. The master layout of the center will strategically correct the dimension, echo to the neighboring complex and form a greenbelt as large as possible at the center of the plot.

Section design

Spatial organization of Kunshan Culture and Art Center adopts the principle of openness, commonness and adaptability to serve the citizens more closely and flexibly . Through careful organization of heights of the section floors and staggered floors, the key function areas and auxiliary function areas of the conference and exhibition center achieve floor-specific layout. The most accessible part at the center is to set as a common area. The common area takes consideration of the possibility for both use requirement and open management to achieve compactness of part closeness and open management. Some frequently used facilities, such as multi-media center and cinema, can be separately opened to the outside. The plane of conference and multi-functional hall considers the possibility for free separation of conference, show, exhibition, banquet and events of different kinds. Multivariant floors inside the building complex and elevation system at the site form many common streamlines of rich experience and encourage the citizens to approach, stop and walk through the buildings.

Elevation form

In line with the internal function attribute, the conference and exhibition area of the center looks like a floating metal box. Materials of the elevation is perforated metal and glass; the theater looks like a statue of white curve mass. The two sections form a dramatic parallel. In the proposal, though without uniform formal logics, the integral form of the center emphasizes integration with integral space of the plot. The idea provides a special perspective to the subsequent proposal.

设计方案3
设计公司：
中国建筑设计研究院（集团）建筑设计总院崔愷建筑设计工作室

Design Proposal 3
Designer:
China Architecture Design & Research Group, Cui Kai Architecture Design Studio

工程概况及区位特点

项目用地位于城市森林公园南侧、昆山市体育中心区西南部，环境优美、交通便利。森林公园以水景和绿化为主，北有商业中心区和文化中心区，南有体育中心区。文化艺术中心用地与体育中心相邻，用地南侧为城市主要干道前进西路，西侧为城市次干道，东侧和北侧为体育中心区内部道路。

与环境相融合的总体布局

昆山市文化艺术中心由1400座的大剧场、会议中心、展示中心、文化活动中心、影视中心及配套商业、车库所组成。项目基地生态自然环境优越，基地内有集中绿化带以及呈曲线形态贯穿南北的水系，是北部森林公园的延续。基于以上现状，如果采用直线式的布局将会与周边曲线形态的自然环境产生明显冲突，破坏景观环境的整体感和延续感。因此，总体布局采用与周边环境相契合的曲线式布局，由南侧逐渐旋转发散，沿水体向北舒展放开。这种契合使建筑与环境相互呼应，增强城市景观的延续性。

创造滨水城市空间

文化艺术中心是一个城市公园，公园内建筑与城市中大多数建筑不同：一般位于城市道路两侧的建筑都贴邻道路布置，而公园中的文化建筑则应位于公园环境之中而不是道路旁。另外，公园离不开水。水体对于昆山来说其重要意义已经上升到了文化层面。水是昆山的灵魂，昆山有三分之一的市域面积被水面覆盖。为此，文化艺术中心的主体建筑被放置于绿化景观环境之中，紧邻水体布置。在滨水的一侧，配备层次丰富的室外平台、休息厅、廊桥、散步道、市民广场和亲水空间，使建筑滨水的一侧成为市民休闲娱乐的开放性场所，让人们在现代建筑中感受到昆山无处不在的水乡特色。

缝合与串联建筑的廊桥

由于种种原因，地段外部现已建成的昆山主体育场和网球馆各自孤立，相互之间缺乏空间上的联系；受到城市主、次干道的隔断，文化艺术中心用地通向西侧和南侧

对页：鸟瞰效果图
Opposite page: Bird's eye view rendering
本页：规划总平面
This page: Master plan

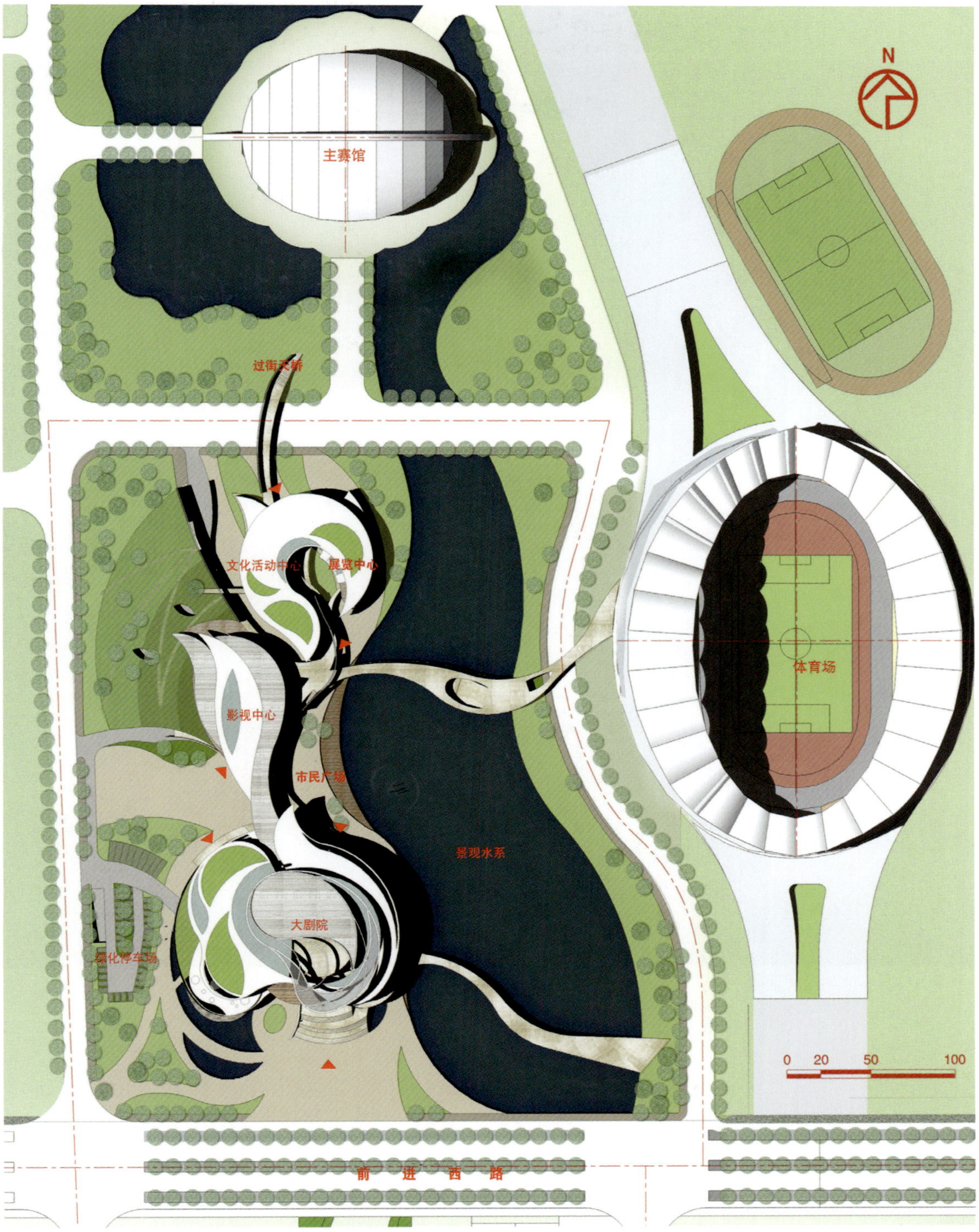

城市街区的步行通道也不畅通。为了增强单体建筑之间的联系，提高各自的可达性，并将若干单体联系成一组群体，在建筑中不同标高上设有一系列自由伸展的曲线形廊桥。廊桥将公众从地段西部入口、北部会展中心入口、东部市民广场、南部前进西路入口等处引入整个文化活动中心，贯穿建筑群体并通向相邻的体育场和网球馆。廊桥为游客提供遮风避雨的室外公共空间，同时也将单体建筑与主要人流集散地紧密地联系起来，将分散的功能块缝合成一个区域性的整体，是强化文化艺术中心整体形象的纽带。

空间意象和形态意象

设计中抽取了具有水乡特色的元素，并将这些传统元素应用到现代的建筑空间中。整个建筑群沿水体呈曲线形布置，并形成自然的进退关系，布置滨水游廊、曲线的驳岸、亲水平台、廊桥等，使建筑具有水乡的"神韵"。

在形态设计上，选取最能代表昆山文化的昆曲和并蒂莲作为母题。昆曲是昆山文化极具代表性的艺术形式，并蒂莲是昆山园艺文化的代表，而地方戏曲和园艺正是广大市民所能广泛接触到的最具渗透力的文化象征。整个建筑的平面形态由昆剧表演艺术中的甩袖形象衍生而成，建筑从一个中心逐渐旋转发散，正如昆曲表演中轻摇手臂翩翩舞动的长袖；而发散出的一簇花瓣的形态恰似盛开的莲花，这又同时呼应了并蒂莲这一母题。

简洁灵动的立面风格

在立面设计中，并没有采用过多结构复杂、建造昂贵的设计元素，而是以错落的平台和楼板为本，以流线型为基础，以白色为主色调，将混凝土、玻璃、穿孔金属板、格栅、木板等材料有机组合，使得建筑凹凸有致，风格简洁明快。整体效果宛如风行水上，酣畅灵动，反映出灵秀江南的传统建筑特点和昆曲水袖的神韵。简约现代的穿孔金属板，利用孔洞的变化调节外立面的透光率，并形成通风的腔体，改善建筑物的外观效果和内部舒适程度，形成疏密有致、玲珑朦胧的独特肌理。屋面部分为室外平台，另有部分采用种植屋面，在产生良好的生态节能效果的同时，也丰富了城市建筑的第五立面。

Engineering overview and regional features

The site used by the project is located on the south side of the city forest park and the southwest of Kunshan gymnastic center. The environment is beautiful and the traffic is convenient. The forest park mainly comprises waterscapes and greening with commercial center and culture center in north as well as gymnastic center in south. The landscape green land used by Kunshan Culture and Art Center is contiguous to the gymnastic center, and the south side of the land is Qianjin West Road which is the city trunk road as well as the city secondary trunk road in west and the interior roads of the gymnastic center of the east and north sides.

Overall layout combined with the natural environment

Kunshan Culture and Art Center comprises the theater with 1,400 seats, the meeting center, the exhibition center, the culture activity center, media center and matched commercial facilities and garages. The ecological natural environment of the project base is outstanding. There are concentrated greening zones and water system running south-north in a curve shape in the base, and it is the extension of the forest park in north. Based on the above conditions, if the layout is linear, it will be obviously conflicted with the natura environment around which is in a curve shape, and the integral and extensive senses of the landscape environment will be damaged. Thus, the overall layout adopts the curve-type layout which fits the peripheral environment. It gradually rotates and extends from the south side, and stretches towards north along the water system. This cooperation makes the building and environment mutually correspond so as to reinforce the continuity of the city landscape.

Creating a space of waterfront city

Kunshan Culture and Art Center is a city park whose buildings are different from most of buildings in the city. Generally, the buildings on both sides of the city road are arranged closely to the road but the culture buildings in the park will be located in the park and not the sides of the road. In addition, water is necessary to the park. To Kunshan, the significant meaning of the water body has been ascended to the culture aspect. Water is the soul of Kunshan. One third of the city area of Kunshan is covered by water. Thus, the main buildings of the culture artistic center are located in the greening landscape in the way of being close to the water body. At the waterfront side, outdoor platforms, rest halls, gallery bridges, walkways, civil squares and waterfront accessible spaces are arranged so that the waterfront side of the buildings becomes the open place for citizens, and people will feel the water features of Kunshan everywhere among the modern buildings.

Connecting and cascading buildings

For various reasons, the main stadium and the tennis hall of Kunshan which have been completed outside the section are respectively separated, and it lacks of spatial connection. Isolated by the main and secondary city trunk roads, the walkways of the land used by Kunshan Culture and Art Center to the west and south city avenues are not convenient. To reinforce the connection among individual buildings, improve

对页：大剧院夜景效果图
Opposite page: Rendering-night of the grand theatre
本页，从左至右：大剧院走廊室内效果图；大剧院入口效果图
This page, from left to right: Rendering of indoor corridor in theatre; Rendering of the entrance of theatre

the respective accessibility and connect a plurality of individuals into one group, a series of curve-shaped gallery bridges which can freely extend are built at different elevations in the buildings. The gallery bridges lead the public into the whole culture activity center from the west entrance of the section, the entrance of exhibition center in north, the citizen square in east, the entrance of Qianjin West Road in south, etc., and run through building groups and reaches contiguous stadiums and tennis halls. The gallery bridges provide outdoor public shelters from the storm for tourists, tightly connect individual buildings with distribution points of main human flow conbine dispersed functional blocks into a regional whole body, and is the link for reinforcing the comprehensive image of the Kunshan Culture and Art Center.

Spatial image and form

For design, the elements of water features are extracted, and these traditional elements are applied to the modern architecture space. The whole architecture group is arranged in a curve shape along the water body, and forms a natural backward and forward moving relationship. Waterfront tourist galleries, curved bulkheads, waterfront platforms, gallery bridges, etc., are arranged to make the building have more essence of the waterfront city.

To form design, the Kunqu opera and twin lotus which can essentially represent the culture of Kunshan are selected as motifs. Kunqu opera is the artistic form which is the best representative for Kunshan culture, and the twin lotus is the representative of the gardening culture of Kunshan while local opera and gardening are the very culture symbols which can be widely contacted by citizens and have highest penetrability. The planar form of the whole building is derived from the sleeve swinging image in the performance art of Kunqu opera. The buildings gradually rotates and spreads from one center, and it is just like the performance of Kunqu Opera in which arms lightly swing long sleeves; and the petals formed by spreading is just like the blooming lotus, which simultaneously reflects the motif of the twin lotus.

Simple and flexible vertical surface

Not many design elements with complicated structures and high construction cost are used in the design of vertical surfaces but it is based on platforms and floor slabs, takes white as the main color tone, and organically combines concrete, glass, punched metal plates, racks, wooden plates and other materials to make the buildings be regularly concaved and convexed and the style be simple and clear. The integral effect is just like a wind runs on water, and it is free and flexible so as to reflect the traditional architecture characteristics of the beautiful regions south of the Yangtze River and the spirit of the water sleeve. To the simple and modern punched metal plates, the variation of holes is used for regulating the light transmittance of the outer vertical surface, and a ventilative chamber is formed. The appearance effect and the interior comfortable degree of the building are improved to form the unique texture which has regular density which is smart. The roof part is an outdoor platform, and some roofs for planting are also available. It generates excellent ecological energy-saving effect and enriches the 5th vertical surface of city architecture.

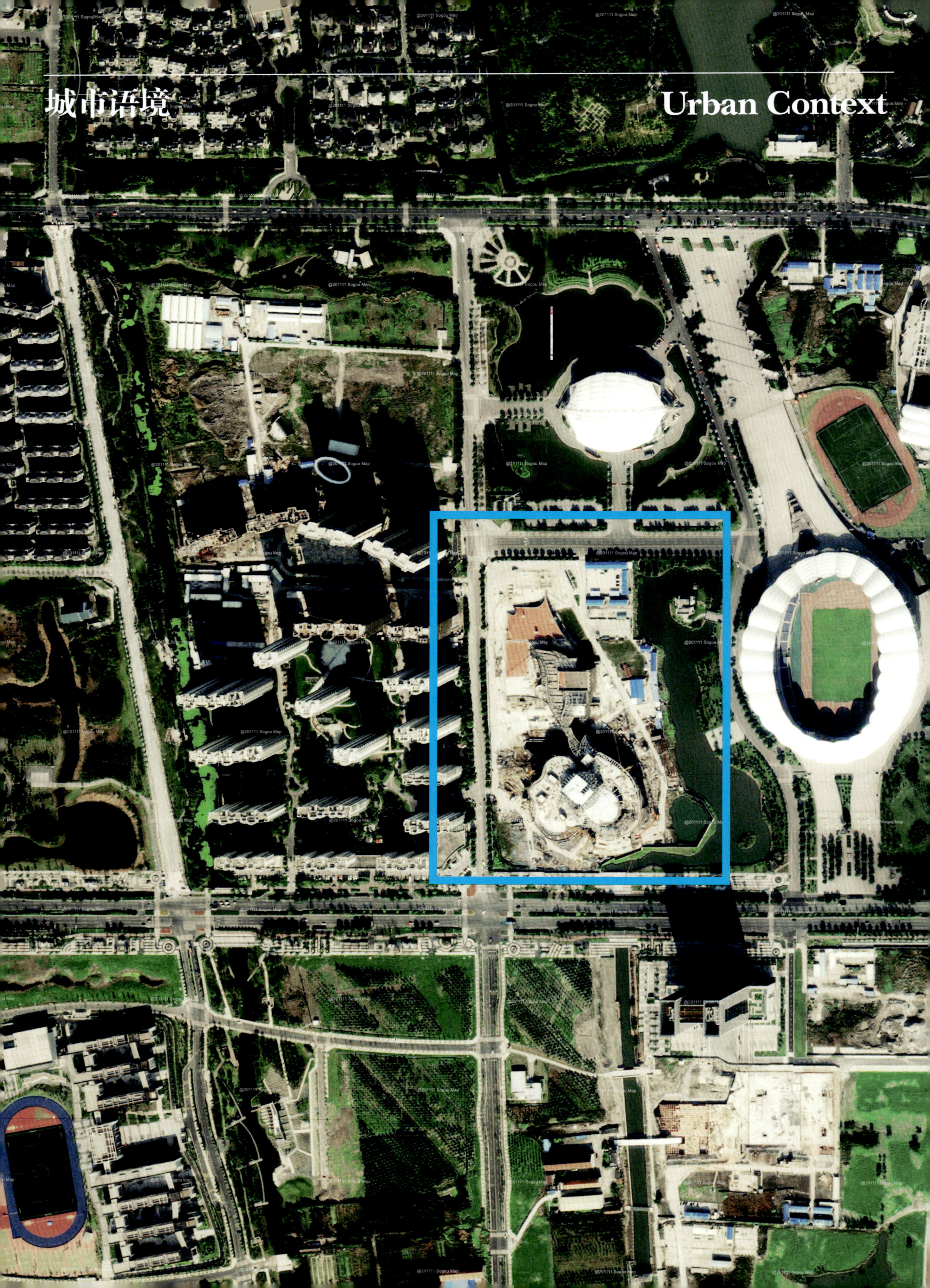

地理位置 Location
中国，江苏省，
昆山市，前进西路1850号
No.1850
Qianjin West Road,
Kunshan,
Jiangsu, China

大剧院以曲为形，以流畅的形态让建筑融入环境，并与体育场馆对话

THE THEATRE TAKES CURVE AS THE SHAPE, BLENDS THE SMOOTH FORM INTO THE BUILDING AND EXISTS IN HARMONY WITH THE SPORTS CENTER

对页：市民广场上的绿化及铺装。
Opposite page: The virescence and decoration on civil square.

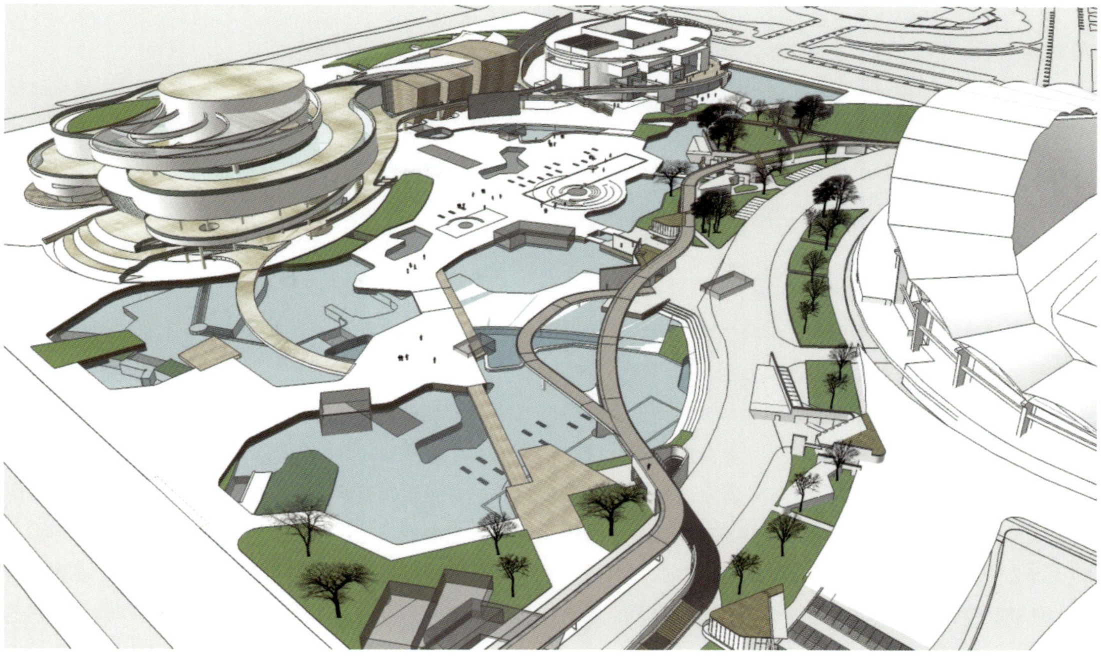

昆山文化艺术中心模型分析图
Analysis diagram of Kunshan Culture and Art Center

总述

为一个地方设计一座大剧院，就好像为这个地方咏唱一首赞美的歌。五年前，当我应邀为昆山设计大剧院时，心中自然萌生这样一种情感，并以这种情感的冲动去看待这片土地，孕育它的未来。

位于阳澄湖畔的这个朝气勃勃的城市原本也是江南水乡，只可惜，在快速城市化中它的那些老街老巷消失了，城市文化也伴随着城市特色的衰微而日渐低落，建设大剧院就是为了重振文化，追求特色，让昆曲的故乡更文脉绵长。

昆山大剧院选址在老城的西部新区，旁边是已建成的体育中心。用地中间有一条小河，北面连接万亩森林公园，南边不远处是娄江，周围是城市道路和正在建设中的新城高层住宅和办公建筑群。

昆山大剧院功能包括一个1400座大剧场，一个500座多功能小剧场。还有由许多大小不同会议厅、室组成的会议中心，另外还有八个影厅组成的电影院，地下还有设备机房和停车库等配套设施。二期还将增建城市规划馆，总建筑面积71255.8m²。

• 布局

这个重要的城市公共建筑在特定的环境中建设首先碰到的是布局问题，布局的策略不同会产生不同的城市空间关系，也影响建筑设计的方向。实际上之前也有其他设计单位做过方案，但也都带来一些问题：或过于集中孤立，没有充分利用场地的资源；或重视了街道界面，却忽视了河道景观带，与体育场馆的关系也比较疏离。经过分析，我们选择了另一路径，把用地与北面的森林公园联系起来当作连续的城市绿带公园来统一考虑，而将剧院作为公园里的建筑，与河道发生更密切的关系，强调它的滨水性，也试图建立与体育场馆更积极的空间联系，在尺度上，色彩上也应与之协调对话。

于是我们将建筑体量依功能沿河边分组展开，各组之间用敞廊和棚架联系，廊道亦向体育场馆方向延伸，意在打破用地范围的局限，让大剧院与体育中心携起手来，共同营造一个开放式的城市滨河景观空间。在沿街一面重点采用绿化景观设计手段，或起草坡、或植竹林、或筑水池、或造小景，把广场和停车场分散掩映在绿荫之中，形成较为自然的公园氛围，实现了“房在园中”的整体意图，也避免出现一般剧院由于正面和背面功能不同而造成的空间界面的不连续的问题。

• 形态

说到建筑形态，一般应该遵从由内到外的原则，所谓“形式追随功能”。但对一个处于城市景观中的建筑，它的外部形态的景观性，也自然成了另一种“功能”，实际上国内外很多造型优美独特的地标式建筑作品，虽然并非是因内部功能所致，却也能受到社会广泛认可，甚至成为城市的名片。可见道理并不绝对，关键看这个形是“徒有其表”呢，还是贯穿内外与功能空间巧妙地组织在一起，达到“一气呵成”的效果。

坦率地说，昆山大剧院的建筑形态的确是由表及里产生的。我们通过对场地性质的分析，采用了建筑在公园中的总体策略，所以建筑形态就在园林景观的外在视野中综合考虑了。首先是亲水，我们希望依这条不宽的河道营造优美的滨水空间，让建筑面水一侧应该有开放性，并借鉴江南水乡沿河设长廊的经典手法，在建筑底层架设外廊、平台。而这外廊亦延长伸展，连接到河对面原本比较消极的体育中心室外滨水空间，目的是带动或激活那部分的城市空间,形成互动的整体。其次是“曲”，场地中弯曲的河道，弧面的场馆，都是既有的要素，所以大剧院也以曲为形，以流畅的形态让建筑融入环境，并与体育场馆对话，这曲墙引自河边，分层缠绕，漂浮游动，蜿蜒伸展进入室内大厅，再翻卷升腾，最后在观众厅顶上缠绕成结，叠合如绽放的玫瑰花瓣达到高潮。第三个关键词是“柔”，江南水乡粉墙黛瓦如水墨画卷轻勾淡染，更有昆曲艺术曲调悠远，水袖飞舞，让我们体会一种轻柔之感。于是我们选用白色铝板幕墙，在建筑的外围缠绕漂浮，墙板盖顶采用半圆形断面过渡，清除硬边，屋面上还有曲梁构成的遮阳花架，电影院顶又有三层薄板叠错，这一系列手段都是“柔化”建筑。在阳光照射下曲面呈影渐浓渐淡，加之穿孔的疏密和大小变化，让其从三维的沉重实体建筑向二维轻薄画卷的意象转化，从而呈现出一种“诗意”。第四个词是“透”，它故然依赖于大厅的超白玻璃幕墙使空间通透，内外一体，让曲墙连续，其实金属幕墙上的穿孔更是一种有趣的微透，白天从内看外自然光穿孔而进，呈光栅般斑斑驳驳，夜晚室内灯光透过微孔呈现一种绸幔透光的效果，十分优雅微妙，是我们特别喜爱的一种意境。

• 技术

无论多么炫目浪漫的建筑构想，在其背后都需要十分具体和复杂的技术支撑，甚至往往越富于动感的建筑对技术的挑战性就越大。

首先是设计技术，要重点解决复杂的曲面体的定位准确，在设计全过程中一直采用三维模型软件进行空间定位和形体控制，建筑和室内设计都在同一套模型中去推敲方案，结构设计和幕墙设计，以至机电设计也都要在三维模型中来表达和检验。

然后是结构技术，多种结构体系混搭是这里的技术难点，坚固的钢筋混凝土框架剪力墙体系支撑建筑功能体的大部分空间荷载，顶部的曲梁悬板和端部的“海角”空间体就要用钢结构来表现它的轻巧飘逸，曲面的悬挑外墙靠的是钢管混凝土圆柱支撑，舒展的外廊用的是单列混凝土圆柱挑梁承担。

项目中幕墙技术包括金属穿孔板幕墙、玻璃幕墙、陶板幕墙。幕墙的图案设计繁琐而有趣。在搜集了大量的昆曲剧照的基础上进行一系列处理，包括筛选、定样、排序、剪辑、拼接等等，将原本相互独立的、静止的一幅幅画面，编辑成连续的、舞动的全屏场景。随后，把色彩关系转化成素描关系，又将黑白灰像素化成4-5个色阶的图案。最后，让不同色阶对应尺寸不同的孔洞，穿孔的大与小、密与疏，还控制着室内透光的多与少、明与暗。随光影变化的金属穿孔板幕墙，让人们在传统与现代之间寻找文化的脉络。在节点构造的研究中，经过多次制作样板进行分析比较，最终两片铝单板冲孔，合成为有密肋支撑的60mm厚的双层空腔板，整个金属幕墙没有横缝，组成装饰图案的孔可快速冲切而成，加工组装便捷高效，铝板和铝型材的连接通过铆钉铆接在一起，使型材与铝板组成的整体将共同成为受力构件，使金属幕墙结构更轻巧，受力更合理，造价也会降低。而铆钉工艺又采用了埋头铆钉技术，让铆接不留痕迹，形成美观的效果，成为建筑立面上的最大亮点受到各方的肯定。玻璃幕墙主要是采用了超白玻璃，最大限度地保持室内外的通透性和整体感。幕墙的玻璃肋加点式驳节点体系，在保证了它的强度和稳定性之外，也保证了幕墙整体的晶莹和雅致。

大厅的地面选材经过多方案比较试制，选用了环氧水磨石新工艺，其不仅可以保证十分光洁的表面，也能避免标准水磨石因温度变形产生裂缝，不做或少做分隔条，特别是它的色彩和骨料的丰富性是传统工艺难以比拟的，骨料中可以采用碎玻璃和不锈钢等反光颗粒，让地面闪烁出点点钻石般光芒。

绝大多数剧院的观众厅基本都是对称的。据声学专家讲这主要是为了声场的对称和均衡。但昆山文化艺术中心大剧院的整体形态是强调动感行云流水。而动感的基本特色就是非对称，只有不均衡才能产生动势。所以在观众厅设计中我一直坚持做不对称的楼座组合。这种特别的楼座组合，也会带来座位视线上的问题，设计小组中的建筑师李斌为此做了仔细的计算和推敲，精确的算出每个座位的位置和方向。但在验收时有人提出有几个座位视线不行，他拿出图来一核对，发现就是那九个座位原本不应该放在那个地方，是装修公司增加上去的，这恰恰说明了他工作的精准度。

基于建筑背景，室内设计在本工程中也面临技术的挑战，室内设计的主要空间界面也都是由曲线或曲面构成的。设计将主要空间的界面进行解构，由不同弧度的曲面交叠穿插，使空间形式丰富而有层次。

此种设计类似于空间雕塑的设计，任何视点的交换都会给观者以不同的视觉观感，必须在三维软件上直接生成空间界面，以确保每一个曲面都是优美的，并能在各个视角经得起推敲。另一个问题是如何将三维模型细分，为施工单位提供可以实施的三维定点。

以艺术中心观众厅为例，设计以捧花为设计概念，将空间中的墙面顶面抽象提炼成花瓣与花心的构成要素，而栏板、墙面还是延续飘带的形态通过灯光的映衬形成的色彩晕染强化了大厅的艺术效果。与此同时，在建模的过程中将各个设备（灯光、马道等）及声学设计有机地融入空间中，使各个造型和界面浑然一体。但由于音响扩声设计系统配合滞后，造成扩声器位置没有很好地融入界面设计中，许多内置和外挂的设备对界面的完整性造成了不利的影响，是一个需要整改的遗憾。

• 合作

昆山艺术中心功能综合，技术复杂，完全是一个团队合作的效果。在设计和施工的全过程中，一直有一种积极合作，各方配合的工作关系和氛围。

首先感谢昆山城投公司周继春董事长对设计师的尊重，项目总经理范晓玲对工作的认真负责使我们对工程的设计一直处于有效并有序的控制之中。

然后应该感谢我们的设计团队，相关建筑、结构、机电、总图、室内设计等各专业的设计师为这个项目团结奋战，付出很多，对工作的认真和解决问题的能力得到了业主和施工各方的认可和好评，最后成功展现的成果也说明了一切，为我们中国院争得了声誉。

另外要特别感谢与我们合作声学设计的张三明教授，他对不对称观众厅的设计挑战很有耐心和自信地应对，积极地提出解决办法，甚至为此做了数字模型和十分之一的大比例的实体仿真模型，使观众厅的声学质量达到了较高的水平。另外法国著名艺术设计师阿兰·博尼先生及其团队认真地倾听建筑师的构想，善于用彩色肌理和灯光去表达艺术的效果，并积极地与室内设计师合作，多次深入现场，选样板，调灯光，让建筑的艺术性有了很大的提高。令人欣慰的是在之后的项目中我们再次携手，开展了新的合作。

还应该提到的是项目中的景观设计师是一位来自香港的青年设计师，据说他在昆山一带曾经做过一些项目，周董对他相当信任和欣赏，应该说他的设计也表现出了相当的水平。意图以丰富的人工水景和地面铺装来打造有活力的城市广场，但在交流中我们提出应整理和利用现状河道景观，多植树和水生植物，软化边界，更应该注意建筑和水的关系，因为这是我们设计的初衷。可惜他以水质不好为由坚持原设计，最终的效果是用大片硬质广场切断了建筑与河道的关系，河道被压缩成很窄，树木亦显得稀疏，背离了总的设计意图，十分遗憾！景观设计师似乎也意识到了这个问题，他希望最终这个广场能够融入城市生活。我也希望如此。

—

文：崔愷、何咏梅、李斌、张晔 中国建筑设计研究院

本页：大剧院和影城屋顶曲线形飘带的搭接。
对页：大剧院休息厅入口。
This page: The curvilinear lapping between theatre and screen center's roof.
Opposite page: The entrance of lounge hall in theatre.

票务中心
大剧院
OPERA

Overview

To design a theater for a place seems like composing a song of praise for the place. I was overwhelmed by such kind of emotion and felt the impulse to treat this land and nurture its future with such emotion five years ago when I was invited to design a theatre for Kunshan.

Located near Yangcheng Lake, this vibrant city used to be a water town. It is such a pity that old lanes and alleys have vanished in the road of rapid urbanization and the urban culture has been casted away with declining urban characteristics. Thus to build the grand theatre is to revive the culture, to pursue urban features and to make the birthplace of Kunqu opera long lasting.

The construction site of the theatre was selected in the western new district in the old town, near the Sports Center that had been built. A river flowed across the

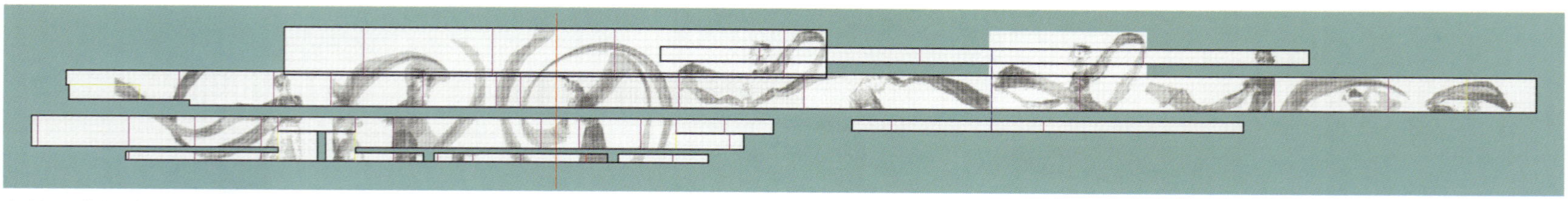

大剧院金属幕墙立面展开图案效果图
Expanding sketch of the metal curtain wall in theater

center of the land which connected with the Forest Park with an area of tens of thousands mu in the north, borders on the Loujiang River in the south and was surrounded by urban roads and new high-rise residences and office buildings under construction.

Kunshan Theatre consists of a large auditorium accommodating 1,400 seats, a small auditorium accommodating 500 seats, a conference center that comprises many conference halls and rooms of various sizes, a movie house including 8 film halls and an equipment house and a parking garage underground and other supporting facilities. The Phase II is designed to build an Urban Planning Museum, with a total construction area of 71,255.8 m^2.

· Layout

To construct such an important urban public building in a particular environment, the first problem encountered is the layout issue, because different layout strategies will not only lead to different urban spatial relationships but also influence the direction of architectural design. Actually, other companies had proposed their designs, all of which brought about some problems. Some design was too centralized, and failed to make full use of the land; some design considered the street layout but ignored the river landscape and to some extent isolated the theatre from the Sports Center. After analysis, we chose another way to design. We unified the land with the Forest Park to the north and considered the park as a continuous greenbelt park of the city and the theatre as architecture in the Park, so that we linked the theatre more closely with the river and emphasized its riverside feature. We also tried to establish a more positive spatial relationship between the building and the Sports Center and blend the two architectures in harmony in both scale and color.

Therefore, according to different functions, we designed the architecture into different groups along the river, all of which were linked by outdoor corridors and canopy bridges. The corridor is extended to the Sports Stadium, intended to break the limitation of land use and blend the theatre with the Sports Center so as to build an open urban riverside landscape. Along the areas near the street, we adopted green landscape designs, like grassy slope, bamboo forest, water pond or small garden landscape, and scattered the square and the parking lot under the green shade of trees to form an atmosphere of natural garden, to realize the overall design of "room in the garden" and to avoid the spatial discontinuity of general theatres due to different functions at the architectural front and at the architectural back.

· Form

Speaking of architectural form, we usually view it

对页上图：大剧院正面入口。
对页下图：在室外连廊后面，影视中心凸出来的多个影厅，好似几艘扬·帆的“木船”。
本页上图：影视中心入口。
Opposite page, above: The front entrance of theatre .
Opposite page, below: Behind the outdoor corridor, several video halls are protruding from media center, look like many "wooden boats" in sail.
Above: The entrance of media center.

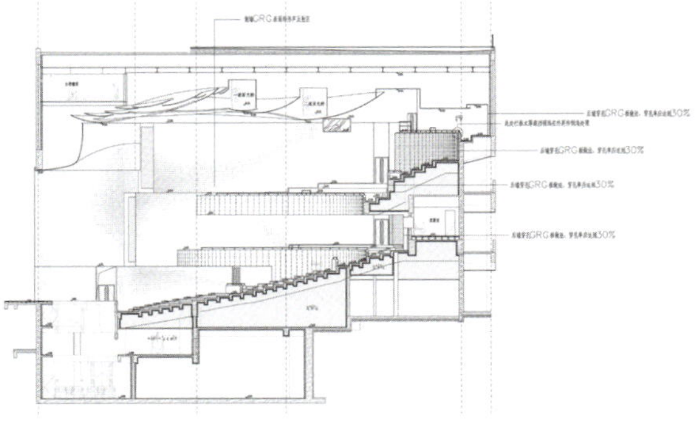

大剧院观众厅东立面图
East elevation of the auditorium in theater

本页:售票厅。
对页上图：影视中心二层连廊。
对页下图：大剧院的室内设计为了强调曲线在空间中舞动的动势，设计将主要空间的界面进行解构，由不同趋势弧度的曲面在交叠穿插中组成空间的各个界面，使空间形式丰富而有层次。
This page: Ticket Room.
Opposite page, above: The corridor on the second floor of the media center.
Opposite page, below: The interior design of theatre aims to emphasize the dancing force of curve in the air. Design deconstructs the interfaces of primary space, constituting the interfaces by the various curved surfaces in trend and radian, thus making the form of space sufficient and clear.

from the inside to the outside, the so-called "form follows function." But for a building in the urban landscape, the scenery of its external shape naturally becomes another "function". Actually many domestic and foreign landmark architectures with elegant and unique form are widely recognized and even become the name cards of their cities, which is however not because of their internal functions. Thus, we can see that truth is not absolute, the key lies in whether the form is just an "outer show" or whether internal and external shapes perfectly integrate with the functional space, achieving the "all in harmony" effects.

Frankly speaking, the architectural form of Kunshan theatre is indeed shaped from the outside to the inside. We analyzed the nature of the construction site, adopted the overall strategy of the building in the park and thus comprehensively considered the architectural form in the external scenery of garden landscape. Firstly, we took the river into consideration and hoped to build a graceful riverside space along the narrow river. The building side facing the river should be open, while the corridor and terrace should be built on the ground according to the classic building style of watertowns that build corridors along rivers. The corridor should extend as far as to link with the outdoor waterfront space of the formerly desolate Sports Center, so as to stimulate and vitalize that part of urban space and form an interactive unity. Secondly, we considered the curve shape. The curving river on the field and the curving stadium were the existing elements, therefore, the theatre should also take the curve shape to blend itself into the environment with smooth form and exist in harmony with the Sports Stadium. The curving walls start from the riverside, diverging, intertwining and winding into the indoor halls and then surging upward to tie a knot on top of the audience hall and overlap upon each other like blooming rose petals. The third key word is about tenderness, we can appreciate the tenderness in the white walls and black tiles in watertowns and from the lingering melody of Kunqu opera and the dancing of long sleeves. Thus we chose the white aluminum curtain wall, adopted semi-circular fault surface to soften and erase hard edges on roof panels and used curved beams on the roof surface as sunshade pergolas. The cinema also had three layers of overlapping thin plates. All of them were meant to soften the architecture. In the sunlight, the curving shade of the architecture waxes and wanes, which coupled with the density and size of perforated shades, changes the heavy 3D physical building into a 2D light scroll painting and exudes a poetic flavor. The fourth key word is "transparency", which means to depend on the ultra-white glass walls to make a transparent space, unifying the inside and the outside to make the curving walls continuous. Actually, the perforated holes on the metal walls show an interesting form of mild transparency. In the daytime, we can see beams of sunlight streaming through the walls from the outside and dancing in the air. During the nighttime, the light inside the building creates a scene of silk mantle by penetrating those holes; presenting an elegant and delicate atmosphere that we are particularly fond of.

· Techniques

No matter how dazzling and romantic architectural ideas are, they all require very specific and complex techniques to be realized. Moreover, the more dynamic the architecture is, the more challenging the techniques will be.

The first is the design techniques. We should focus on solving the accurate orientation of complex curved surface; adopt 3D modeling software to control the spatial orientation and architectural form in the whole design process, refine architectural and interior design in the same set of models and express and test the structural design, curtain wall design and mechanical and electrical design in 3D models.

The second is structure technique. It is a technical difficulty to mix buildings of various structures. The sturdy reinforced concrete frame and shear wall support most functional space. The top hanging plates of curved beams and the end "Cape" space body require the use of steel to show their light and elegant features. The curved cantilever walls should rely on steel pipes and concrete columns for supporting. The outstretching corridors adopt the single-row cylindrical concrete cantilever beams to bear the loading.

The final one is the curtain wall technique. We adopted perforated metal curtain walls, glass walls and ceramic walls. According to different positions, the metal curtain walls are categorized as single panel and double panels, perforated plate and non-perforated plate. The pattern design of curtain wall is both complicated and interesting. After collecting a large number of stage photos of Kunqu opera, we conducted a series of processing, including screening, sampling, sorting, editing, splicing, etc., and edited the originally independent and still pictures into a continuous, dancing full-screen scene and subsequently changed color elements into drawing elements and turned the monochrome into the pattern with 4-5 color tones.

Finally, we matched different tones with holes of different sizes. The perforated holes, large or small, dense or sparse, control the interior lighting, more or less, light or dark. The perforated metal curtain walls change with the light and shadow, so that people can perceive the thread of culture between tradition and modernity. In the study of node structure, after making several models for analysis and comparison, we finally chose two punching aluminum veneers to be synthesized into a 60-mm-thick cavity plate with dense ribs for support. The entire metal walls have no transverse joints; the decorative perforated holes can be quickly punched; the processing and assembling are convenient and efficient; the aluminum plates and materials are connected by rivets, so that the section bars and aluminum plates will work together to serve as the load-supporting components, making the metal curtain wall structure more compact, the load bearing capacity more reasonable and the cost more reduced. The rivets adopt the techniques of countersunk rivets and leave no trace of riveting, achieving the beautiful effects and becoming the biggest bright spot on the building façade which is widely affirmed. The glass curtain wall mainly uses the ultra-white glass to maximize the transparency and sense of wholeness inside and outside the building. The glass ribs of curtain wall adopt the node system to ensure the strength and stability as well as the translucence and elegance of the overall walls.

After trials of various programs, the materials for the hall ground adopted epoxy terrazzo, a new technology, which can not only guarantee a very smooth surface, but also overcome the cracks of deformed standard terrazzo due to temperature. We used no or few separate bars. Especially the richness of color and composition are incomparable to traditional materials, as the composition can include glass fergments and stainless steel and other reflective particles, making the ground flashing and shining like diamonds.

Most auditorium of a theatre is basically symmetrical, which is mainly for the sound symmetry and balance according to acoustics experts. But the overall shape of Kunshan theater emphasizes dynamic, natural and smooth features. The basic characteristic of dynamic beauty is the asymmetry, only to produce uneven momentum. Therefore, I have been adhering to asymmetric combination in the design of auditorium. This particular combination will brings sight problems on the seat. The architect Li Bin of design team made meticulous calculation and deliberation on this aspect and worked out the accurate position and direction of every seat. During the acceptance inspection, someone found that a few seats had poor sight, he checked out the design layout and discovered that the nine seats were not designed in the print but additionally placed by the decoration company, showing architect Li's precision in his work.

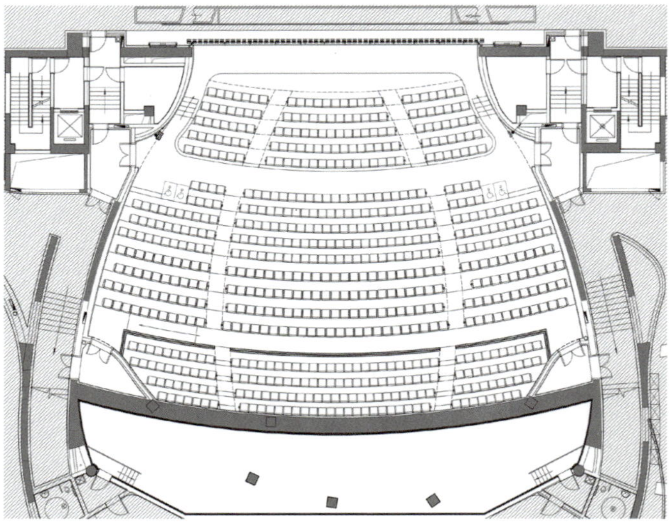

大剧院观众厅一层平面
First floor plan of the auditorium

本页：大剧院前厅连廊。
对页上图：大剧院化妆室。
对页下图：位于大剧院三层的入口之一。
This page: The corridor of front hall in the theater.
Opposite page, above: Dressing room of the theatre.
Opposite page, below: One of the entrances on third floor.

Based on architectural background, the interior design in this project also faced the challenge of technology. The main space of interior design was also formed by a curved lines or surfaces. The design deconstructed the interface of major space which is overlapped by different curved surfaces to make rich and hierarchical spatial form.

This design is similar to space sculpture. Any exchange of viewpoints will give viewers a different visual perception, so that the space interface should be generated directly on 3D software to ensure that every surface is beautiful and every viewpoint stands up to scrutiny. Another issue is about how to subdivide the three-dimensional model and provide applicable 3D fixed points for construction company.

Take the auditorium of Kunshan Culture and Art Center as example. The design of flower concept refines the top surface of wall space into elements of flower petals and pistils, while the breast board and wall surface continue the form of ribbons, which intensifies the artistic effects in the auditorium by hues reflected by light. Meanwhile, all devices (lights, bridleways, etc.) and acoustic designs were blended into the space in the modeling process, integrating all shapes and interfaces as a whole. Due to the lagging sound reinforcement design, the loudspeaker was not well integrated into the interface design. Many internal and external devices caused a negative impact on the wholeness of interface, which is regretfully necessary to be restored.

· Cooperation

The comprehensive functions and complex techniques of Kunshan Culture and Art Center is entirely the outcome of teamwork. In the entire process of design and construction, there was always an atmosphere of positively coordinating working relationship among all members.

We should first thank Zhou Jichun, chairman of Kunshan City Construction Investment & Development Company for his respect for designers and general manager Fan Xiaoling for her responsible working attitude, which enabled us to keep the project design under effective and orderly control.

We should also thank our design team and designers on architecture, structure, electro mechanics, overall layout and interior design and other professional designers for their united efforts on this project. Their serious working attitude and good problem-solving ability are recognized and praised by our proprietor and the construction company. The final outcome of the design gives the most eloquent proof to the ability of the design team, winning great reputation for our institute.

Special thanks should be given to Professor Zhang

Sanming who cooperated with us in acoustic design, responded to design challenges of asymmetrical auditorium with plenty patience and confidence and positively proposed solutions. He even made a digital model and a 1:5 scale large simulated model, upgrading the acoustic quality of the auditorium to a higher level. Famous French art designer Alain Bony and his team engaged in the project in a later time, but they listened to the ideas of architects seriously. They excelled at using color texture and lighting to express artistic effect and actively cooperated with the interior designer and made many on-site studies to choose sample plates and adjust lighting, greatly improving the artistic effect of the architecture and receiving acclaim from the proprietor. It was gratifying that we cooperated again in the Kunshan Cultural Square project and injected new vitality to the project. It should also be mentioned that the landscape designer in the project is a young designer from Hong Kong. He was said to have conducted some projects around Kunshan, so that chairman Zhou showed him considerable degree of trust and appreciation. It should be said that he showed considerable high level in the design. He intended to enrich the artificial water features and floor coverings to build a vibrant city square, but during the communication we proposed to utilize the river landscape, grow more trees and aquatic plants, soften the border and pay attention to the relationship between architecture and river, which was our original design intent. Unfortunately, he insisted on his original design because of poor water quality as he said. It was a pity that the large ground of rigid materials cut off the relations between the architecture and the river, as the river way was narrowed and sparse trees also appeared to be a departure from the overall design intent. The landscape designer seemed to have recognized this problem, and he hoped that eventually this square would be able to integrate into city life. I hope so.

—

Text: Cui Kai, He Yongmei, Li Bin, Zhang Ye of CADRP

景观设计

文化广场是文化艺术中心的重要组成部分，除了为该项目悉心打造场景和满足其相关功能上的要求，同时也定位为昆山西部副中心城市南北轴开放空间系统的关键。

打造的场所既要达到衬托为建筑的背景，也应能独当一面成为市民活动的目的地，并为未来与城市轴在线往南的市民广场和北端的森林公园的动线留下伏线。

为配合场地的综合性使用，调整的河道旁安排了三组约共1730m²的商业配套，丰富岸边空间之余，同时一并与未来城市南北轴架空人行桥的结合考虑，在广场的东界面营造一条三维的景观空间，与配套建筑的屋顶平台结合，更与文化建筑群第二层的飘带概念互相辉映，环抱文化广场。

文化建设的三个联合单体建筑的位置分布主宰了景观布局的策略，两个主要面对城市的界面设在地块的南面和西面，分别是剧院的人行和车行正门。由于地块东面的副地相对较宽(平均约150m)，能够同时观赏到三幢建筑的整体效果，场地的面积也可同时满足创造聚人的场所，因此对场地内原有白塔河水系的调整成为本项目景观层面中最重要的门面。

场地内原有的水体可被视为设计上基本性的约束，但站在昆山建城的层面，以水与城市发展的历史文化关系是不可取代的，因此借用水来编奏该文化广场就更顺理成章。相对而言，场地内水景并非像过往一般单以水景的效果作为主导，更重要兼顾了集水的考虑和调整河道对满足排洪功能等相关策略，使雨水和部分河水通过一系列的水生植栽和承载池来改善水质，再应用于静态水景，既能满足大量用水要求，同时适应昆山降雨量高的情况。

至于动态的旱喷则由东西和南北两条轴线组成。连接地块西入口与东界面的广场中轴通过降低场地高程而增加构建一层顶部空间的高度，并使用连贯性的旱喷水景将两个相隔的空间在视觉上连为一体。文化广场的主要开畅空间平日以三个不锈钢反光半圆体水景作为视线焦点，与剧院的镜面池在概念上具有相同的意义。当启动旱喷时可形成长达80m的戏水空间，锁定空间的界线与纵横交错的水柱又能营造一个活泼而又具丰富变化的脸孔。整套旱喷水景使用多达400个喷头。

场地的主要植栽和光效以相对规整格式布置，提供对应的空间结构和方向感，总体上达到能够抵触建筑群相对较大的尺度和弧线。这些较严谨的景观节奏将随着水景和其光效在晚间散发属于广场更亲切和谐的氛围。

—

文：周保华 香港周保华设计有限公司

Landscaping design

Cultural Square is an important part of Kunshan Culture and Art Center. It aims to meticulously create scene for the cultural construction and meet its relevant function requirement. Meanwhile, it is also positioned as the key of open spatial system along the south-north axle of Kunshan as a western sub-center city. The created place shall not only set off the building but become an independent destination of citizen activities. Meanwhile, it will lay a potential circulation of the citizen square to the south and forest park at the northern end along the city axle in future.

To support comprehensive use of the site, three sets of commercial facilities are arranged along the adjusted riverside, covering about 1,730 m^2, to enrich the waterside space. Meanwhile, overhead pedestrian bridge along the future south-north axle of the city is also taken into consideration. In the east interface of the square a 3D landscape space is created to combine with roof platform of the supporting building, echoing to the ribbon concept of the second floor and surround the cultural square.

According to the strategy, landscape layout is dominated by positions and distribution of the three individual buildings for the cultural construction: two interfaces facing the city are set on in the south and west respectively of the plot as portals of the theater for pedestrians and vehicles respectively. Since the auxiliary plot to the east is relatively wide (averagely about 150 m), one can enjoy the integral effect of the three buildings at the same times. Area of the site can also meet the requirement for creating a gathering place. Therefore, it becomes the most important landscape front of this project to adjust the original Baita River System on the site.

The original water bodies on the site can be deemed as basic binding in the design. On the perspective of urban construction of Kunshan, the historical and cultural relationship between water and urban development is indispensable. As a result, it is more logical to adopt water in creating the cultural square. Comparatively, waterscape on the site is not dominated just by water effect. More importantly, it considers water gathering and waterway adjustment to meet flood discharging function and other relevant strategies. Rain and some river water can be improved through aquatic planting and bearing pool and applied in static waterscape to meet mass water consumption and respond to high rainfall of Kunshan.

The dynamic dry-land fountain is composed of two axles, east to west and north to south. Central axle of the square connecting the west entrance of the plot and the east interface increases the height of roof space of a floor through lowering ground elevation. Continuous dry-land fountain waterscape is adopted to visually integrate two separated spaces. The open space of the cultural square usually adopts three stainless steel light-reflective semi-circular waterscapes as eye-catcher similar to mirror pool of the theater in terms of concept. When the dry-land fountain is started, a water space up to 80 m can be created. The borders defining the space and the crisscrossing water columns can also create a lively and colorful face. The whole set of dry-land fountain waterscape adopts as many as 400 muzzles.

Key plants and lighting effect on the site are arranged by relatively neat format to provide corresponding spatial structure and directions. Generally, it can reach relatively large size and curve offsetting the building complex. These prudent landscape rhythms will give out a more intimate and harmonious atmosphere of the square at night with the waterscape and the lighting effect.

—

Text: ZhouBaohua of FRC design

对页上图：西面入口通过旱喷与光效一直贯穿至东面主广场。

对页下图：驳岸石笼铺砌刻意与建筑的立面创造强烈对比，利用原有的河道将建筑与景观的倒影一起融会在水中。

本页：植栽的丰富层次既要满足首层的直观感知，同时兼顾与第二层平台的对话。

Opposite page, above: By dry spraying and lighting effecting, the entrance of west side penetrates to the main square of east side.

Opposite page, below: Revetment paved with stone gabionand the facade of architecture create a strong contrast, by using the former streamway combining the architecture and reflection of landscape within water.

This page: The bountiful plants not only meet the need of visual perception on the first layer, but also form a dialogue with the second flat as well.

① 大剧院 Theatre
② 影视中心 Media center
③ 城市展览馆 The City Exhibition Hall
④ 白塔河（局部现有河道）Baita River (part of existing riverway)
⑤ 未来城市南北轴架空人行桥 The south-north axle footbridge on stilts of future city
⑥ 大剧院主入口 Main entrance of theatre
⑦ 东西轴下沉部分 Sinken part on east-west axle
⑧ 东入口广场 Square on east entrance
⑨ 水生植栽和承载池 Water plant and bearing pool
⑩主广场（戏水空间）Main square (paddling space)

景观总平面
Landscape site plan

图纸　　Drawings

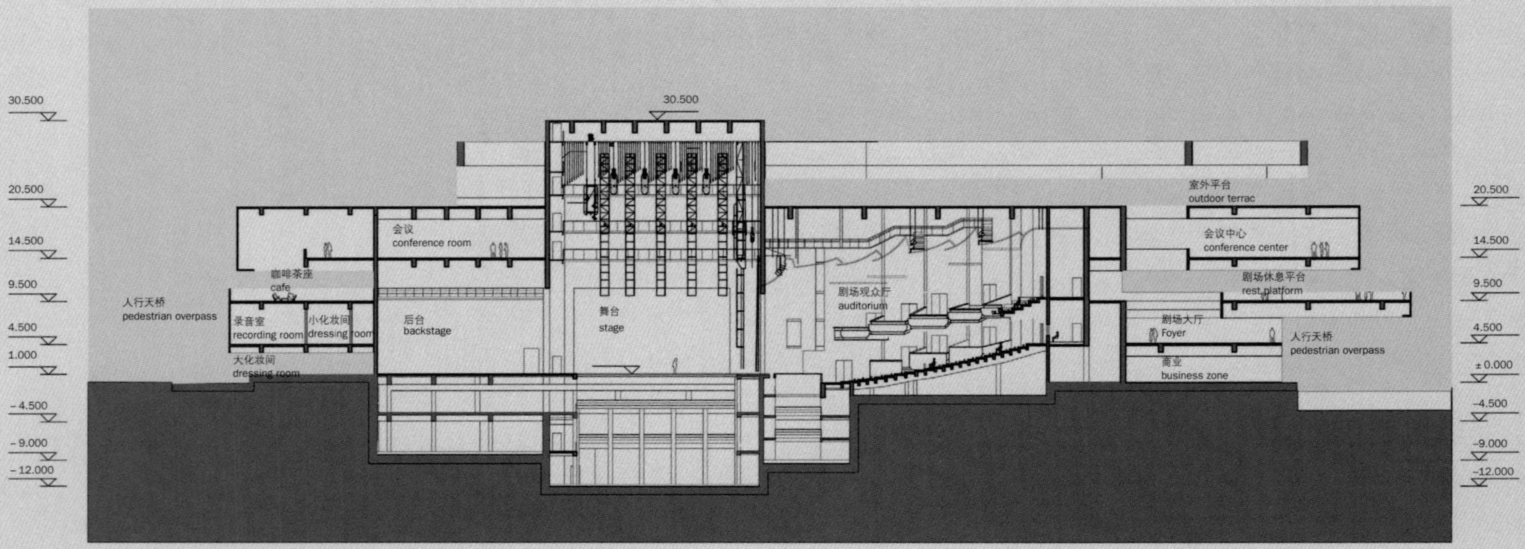

大剧院纵剖面图
Longitudinal section of auditorium

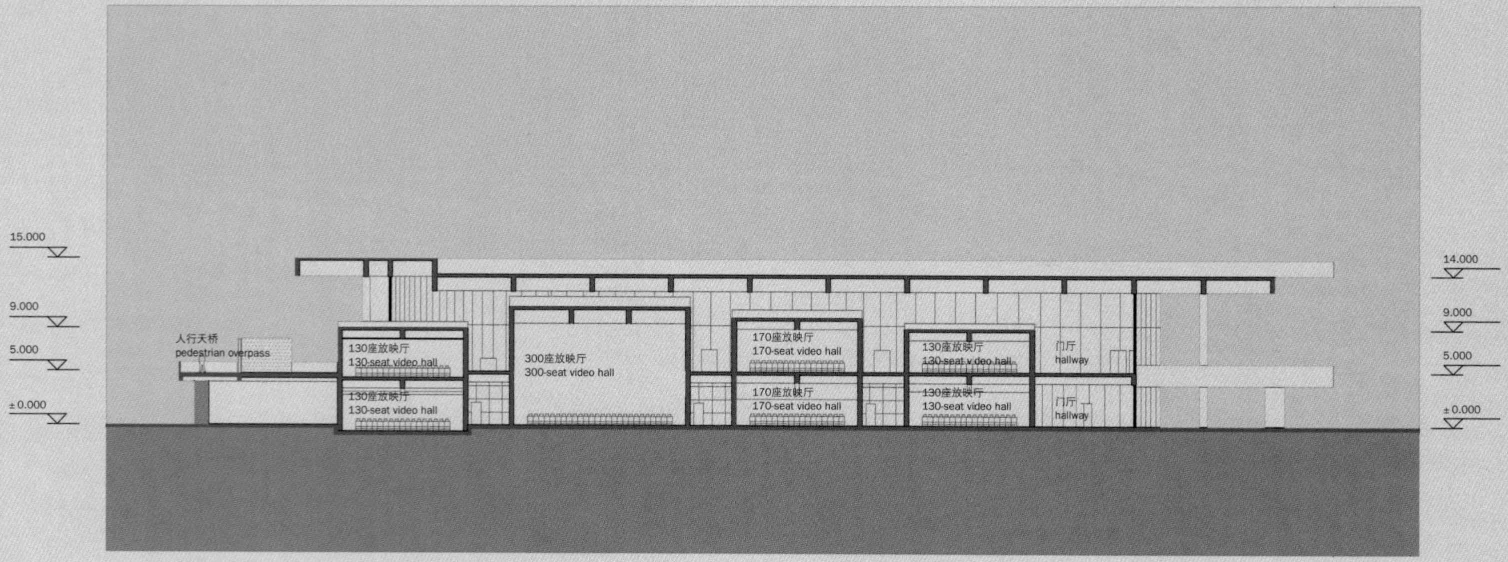

影视中心纵剖面图
Longitudinal section of media center

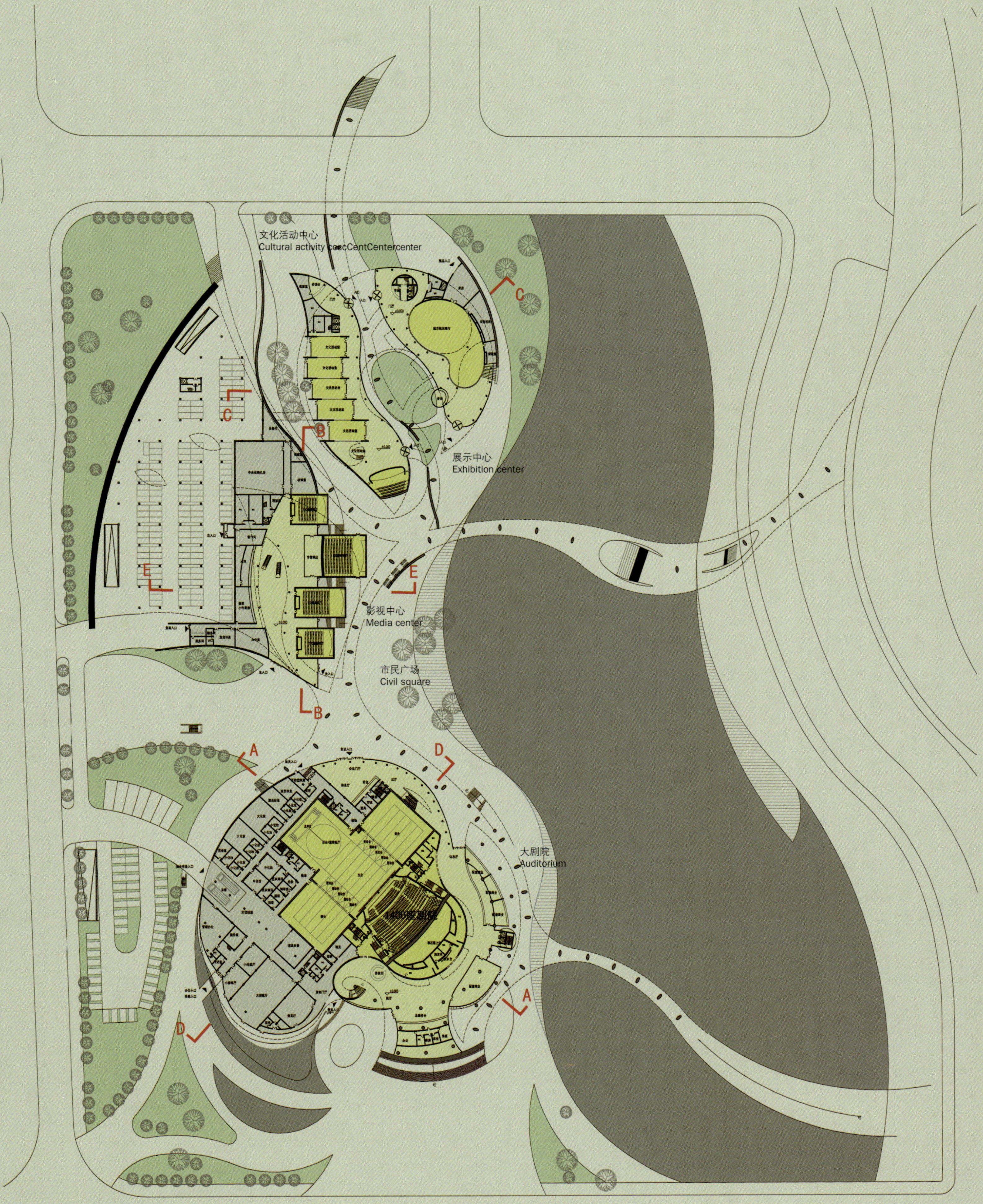

一层组合平面图
First floor plan

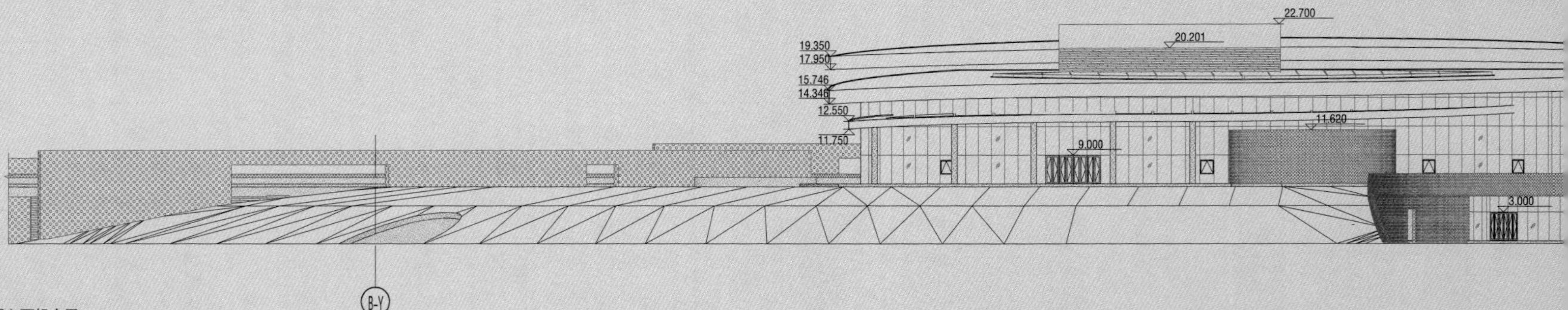

西立面组合图
West elevation

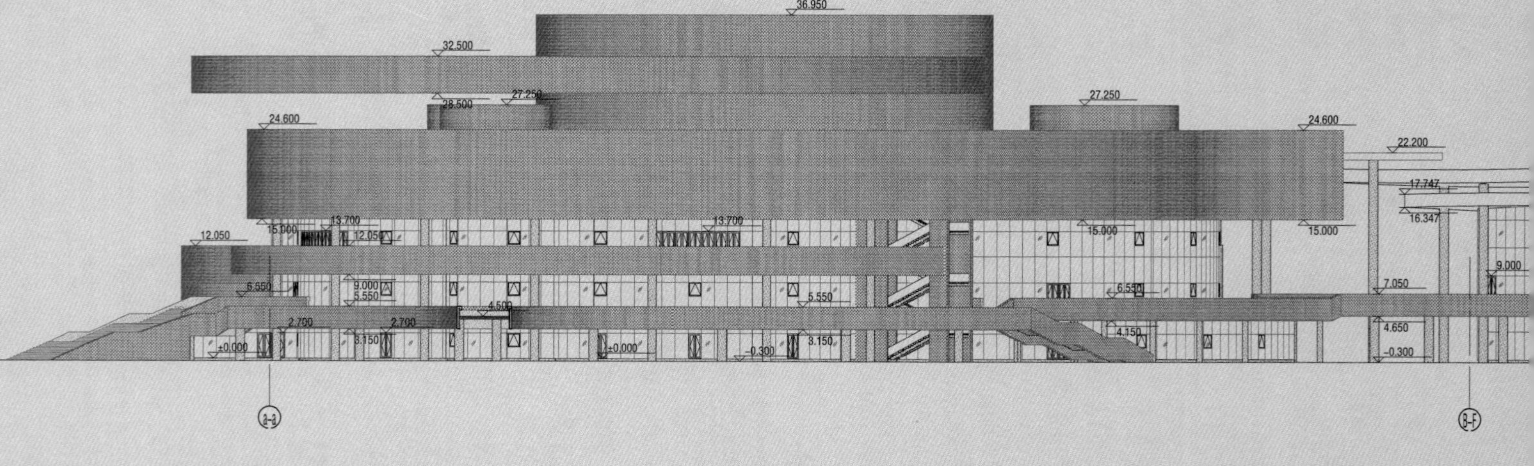

东立面组合图
East elevation

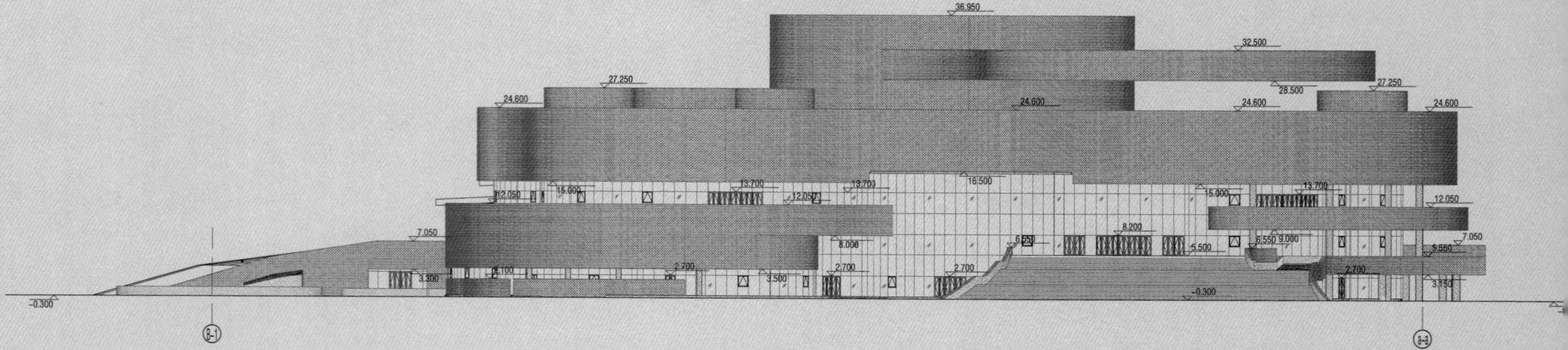

南立面组合图
South elevation

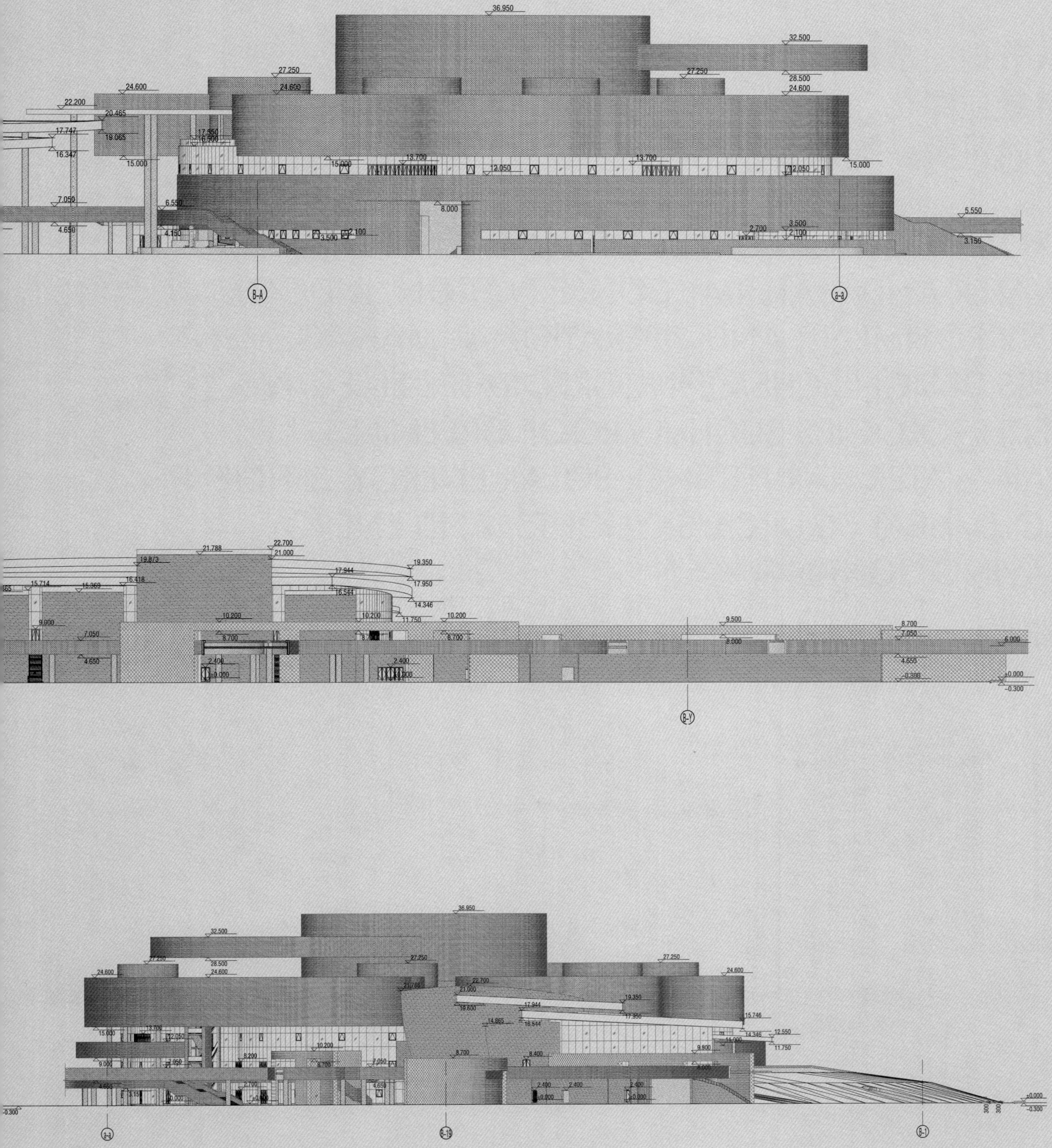

北立面组合图
North elevation

绿色设计 Sustainable Design

从节地、节能、节水、节材、室内环境、运行管理出发，将屋顶绿化、雨水回用、太阳能、高效冷热源、采光顶、节能照明等技术应用其中

FROM THE PERSPECTIVES OF LAND, ENERGY, WATER AND MATERIAL CONSERVATION, INDOOR ENVIRONMENT, AND OPERATIONAL MANAGEMENT, THIS DESIGN SCHEME INCORPORATES EFFECTIVE TECHNOLOGIES SUCH AS ROOF GREENING, RAINWATER HARVESTING, SOLAR ENERGY, EFFICIENT COLD/HEAT SOURCE, SKYLIGHTS, AND ENERGY-SAVING LIGHTING

构思过程

本着对资源节约和有效利用的原则，选用符合昆山地域及建筑特点的绿色生态技术，打造符合绿色建筑二星级要求的文化艺术建筑。

设计理念

将绿色环保与文化气息相结合，将生态技术与功能区域相融合，将美的感知建立在生态影响的基础上。艺术与环保一体化设计，为栖息于文化艺术中心的人们提供一个绿色、节能、舒适而又不失艺术氛围的环境。

设计愿景

在不破坏建筑设计前提下，充分利用各种绿色生态节能技术，保证建筑稳定运行的同时，最大限度地为建筑节省运行能耗。在保证室内外环境的基础上，为游客提供更加清新自然、舒适愉悦的活动空间，同时也为节能减排可持续发展做出一份贡献。将艺术与生态相结合，致力于打造一座绿色生态园式的艺术殿堂。

设计布局

从节地、节能、节水、节材、室内环境、运行管理六大类出发，将屋顶绿化、雨水回用、太阳能、高效冷热源、采光顶、节能照明等技术应用其中，使节能环保、绿色生态的具体要求通过技术手段得到落实，并在满足功能的同时实现宜人生态空间。

节地与室外环境

包括两方面：一方面是绿化配置。在大剧院和影视中心屋顶适合区域设置绿化，选用适宜昆山气候和土壤条件的植物，缓解城市热岛效应，屋顶绿化面积达9396.3m^2。另一方面是透水地面。室外绿化及植草砖为整个场地的雨水自然渗透作出了巨大的贡献，减轻了市政雨水管网的压力。

节能与能源利用

围护结构保温采用陶瓷纤维板、混凝土砌块等保温材料，热工参数满足《公共建筑节能设计标准》（GB50189-2005）要求。

空调系统配置方面，大剧院空调冷源采用水冷冷水机组，影视中心冷热源采用螺杆式地源热泵机组，机组能效比符合现行国家标准《公共建筑节能设计标准》GB 50189中的要求。

而化妆室等小开间人员用房采用新风加风机盘管的系统，新风机组带全热回收装置，减少空调冷热源负荷。夏季节约运行耗电量为25474.5度，冬季节省市政蒸汽291.6吨，年节省费用56119.5元。

观众厅、电影厅等大开间用房采用全空气可变新风比的系统，过渡季全新风运行，利用室外天然冷源消除室内的余热，同时改善室内的空气品质。

节能照明采用高效节能灯具进行室内照明，照明功率密度值不高于现行国家标准《建筑照明设计标准》GB 50034规定的目标值。

太阳能热水系统采用温差强制循环装置制取55℃生活热水，辅助热源采用空气源热泵，为大剧院淋浴提供热水。太阳能提供热水水量为15.5 m^3/d，占总热水量的比例为79%。

屋顶光伏电站共使用180W单晶硅电池组件1140块，

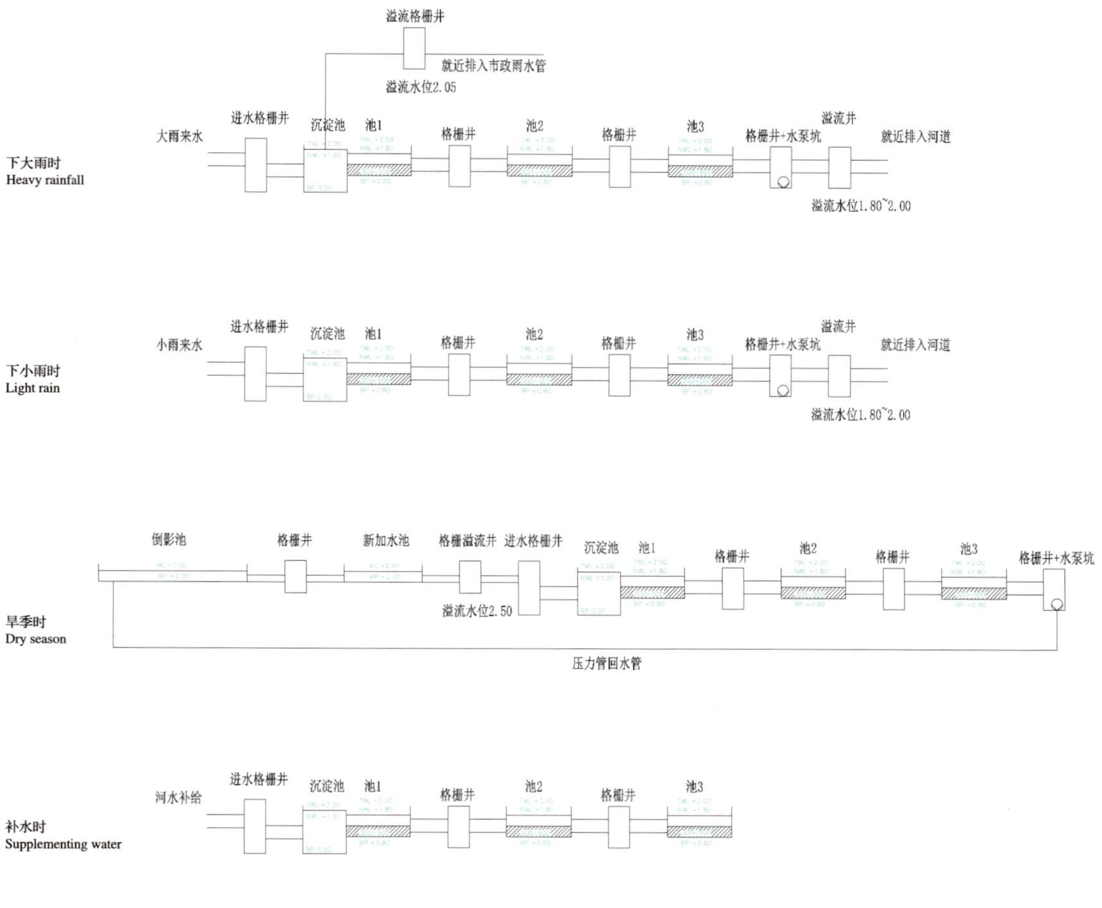

雨水生态处理系统示意图
Conceptual diagram of rainwater treatment system

225W单晶硅BIPV组件34块，整个光伏系统的年发电量为251270.96kWh。

建筑综合能耗采用eQUEST软件对设计建筑与参照建筑进行年耗能量计算。结果如下：

通过加强围护结构保温、采用高效机组、节能灯具等措施，降低建筑能耗。本项目相对参照《公共建筑节能设计标准》GB 50189-2005设置的参照建筑节能14.7 %。

节水与水资源利用

雨水集蓄利用系统结合室外景观设计，雨水采用生态工艺进行处理，沉淀池及两个生态处理池之间存在高度差，场地中上游雨水经过格栅井进入沉淀池，沉淀池中的雨水利用高度差依此进入两个生态处理池，最终流入倒影池，处理后的雨水通过供水泵送至用水点，雨水用于室外景观补水、绿化浇洒及道路冲洗。年利用雨水总量约5661.16 m³。

节水器具卫生洁具及五金配件均采用符合《节水型生活用水器具》（GJ164-2002）标准规定的结局与配件。绿化节水喷灌室外绿化则采用喷灌形式。

节材与材料资源利用

建筑造型要素简约，装饰性构件造价占总造价的比例为3.5‰。并采用可再循环利用材料，占所有建筑材料总重量的比例为 12.8 %。

室内环境质量

大剧院和影视中心屋顶设置采光顶，改善相应处室内休息厅、会议室及门厅的室内自然采光，提高建筑整体的采光效果。经模拟：整体主要功能空间约有75.81 %的采光系数达到《建筑采光设计标准》GB 50033—2001相关功能房间采光系数的要求。

建筑周围立面的通风口分布均匀，室内中庭分布较多，有利于室内自然通风。在过渡季、冬季主导风向平均风速边界条件下，各层室内主要功能空间的换气效率均在2次/h以上。

运营管理

建筑智能化系统合理定位，信息网络系统功能完善。

沟通合作

昆山城市建设投资发展有限公司聘请中国建筑科学研究院上海分院作为绿色咨询单位，设计之初，多次召开会议，与中国建筑设计研究院的设计师们沟通，提出合理性的绿色建筑设计方案。施工过程中，多次亲临现场与施工方沟通，提供技术支持。各方积极配合，最终，该项目成功获得了绿色建筑二星级认证。

—

文：邵文晞、邵怡、韩晶行 中国建筑科学研究院上海分院

本页及对页：在室外景观设计中，室外绿化及植草砖为整个场地的雨水自然渗透做出了巨大的贡献，减轻了市政雨水管网的压力。

This page and opposite page: In the design of landscape, outdoor greening and planting straw-bale make great contributions to the entire field of rain infiltration naturally, with the ease of pressure on the municipal rainwater pipe network.

Design ideation

To uphold the principle of conserving and effectively utilizing resources, this scheme selects green and ecological technologies that suit the unique geographical and architectural characteristics of Kunshan, in an effort to create a cultural and arts building that meets the requirements of two-star green building.

Design concept

By adopting a design scheme integrating art and ecology, the design is aiming to provide people with a green, energy-conserving, comfortable and artistic atmosphere, so as to combine environmental protection and cultural cultivation, incorporate ecological technologies into function zones, and realize aesthetic perception upon ecological consciousness.

Design vision

Various green, ecological energy-conserving technologies are fully made use of without compromising the architectural design, so as to ensure the building operates in the smoothest and most energy-efficient manner possible; on the basis of good indoor and outdoor environments, a more refreshing, natural, comfortable and pleasant space is created to provide visitors for conducting all kinds of activities, while also striving to achieve energy-saving and emission reduction to contribute to sustainable development. By combining art and ecology, this design is to build an outstanding venue of arts which is also green and ecologically friendly.

Design strategy

From the perspectives of land, energy, water and material conservation, indoor environment, and operational management, this design scheme incorporates such effective technologies as roof greening, rainwater harvesting, solar energy, efficient cold/heat source, skylights, and energy-saving lighting, so as to ensure that the specific requirements for energy efficiency, environmental protection and ecological preservation will be sufficiently satisfied, to build a people-friendly ecological space that can meet all relevant functions.

Land efficiency and outdoor environment

The design for land efficiency and outdoor environment includes two aspects. Firstly, greening arrangement: to grow plants suitable for the climate and soil conditions unique to Kunshan in proper areas on the rooftops of the auditorium and the media center, so as to reduce urban heat island effect, with the total area of green roofs amounting to 9,396.3 m^2; secondly, permeable ground: outdoor greening and planting straw-bale make great contributions to the entire field of rain infiltration naturally, with the ease of pressure on the municipal rainwater pipe network.

Energy efficiency and energy use

Ceramic fiber boards, concrete blocks and other insulation materials are adopted for insulation of building envelop, with relevant thermal parameters conforming to the requirements of National Standard GB50189-2005.

Water-cooled chillers are used as the cold source for the auditorium' s air-conditioning system, while screw-type ground source heat pumps are used as the cold/heat source for the media center, with their energy efficiency ratio conforming to relevant stipulations of National Standard GB 50189.

For small spaces such as dressing rooms, the new air+ FCU (Fan Coil Units) system is adopted, with the new air unit equipped with a full heat recovery device to reduce cooling and heating load, thereby cutting summer electricity consumption by 25,474.5 kilowatts and winter steam consumption by 291.6 tons, slashing annual energy bill by RMB 56,119.5 Yuan.

For larger spaces such as the auditorium and cinema hall, designers adopt a full-air air-conditioning system with a variable fresh air ratio, which operates in a fresh-air mode during transitional seasons, capable of using outdoor natural cold sources to reduce indoor heat while improving the indoor air quality.

To achieve energy-efficient lighting, energy-efficient light bulbs are used for indoor lighting, with their lighting power density no higher than the target value as specified by National Standard GB 50034.

A forced circulation device for solar water heating system is utilized to produce 55 ℃ domestic hot water, with air source heat pump as supplementary heat source, which can provide hot shower water for the auditorium. The solar hot water volume provided reaches 15.5m^3/d, accounting for 79% of the total hot water.

Rooftop solar power stations use 1,140 BIPV modules of 180W Single-Crystalline Silicon and 34 BIPV modules of 225W Single-Crystalline Silicon, with the annual

electricity production of the entire PV system accounting to 251,270.96 kWh.

By taking measures of strengthening envelope insulation, adopting high-efficiency units, and using energy-saving light bulbs, we have effectively reduced building energy consumption. Compared with relevant criteria specified in National Standard GB 50189-2005, this project has achieved an impressive 14.7% reduction in energy consumption.

Water conservation and utilization of water resources

The project's rainwater harvesting system integrates with the outdoor landscape treating rainwater by ecological techniques. As there are height differences among the sedimentation tank and two treating ponds, upstream rainwater will flow through the grill well into the sedimentation tank, from which the accumulated rainwater will further flow into two treatment ponds due to height differences and finally stream into the mirror pool. Water supply pumps will then transport the treated rainwater to water consuming point, mainly for the purpose of water replenishment to outdoor landscape, greening and roads washing. The project' s annually harvested rainwater totals 5,661.16 m^3.

本页：太阳能光伏发电电池板。根据测试结果，昆山属于太阳能资源四类地区，境内太阳辐射年总量为12479J/m^2，太阳能热水系统采用温差强制循环装置制取55℃生活热水，辅助热源采用空气源热泵，太阳能提供热水水量为15.5 m^3/d，占总热水量的比例为79%。

跨页：场地水系。

This page: Solar photovoltaic system. According to test results, Kunshan belongs to the fourth level region of solar energy resources, its solar radiation within the area is a total of 12,479 j/m^2. Solar water heating system uses forced circulation device for 55 ℃ domestic hot water and air source heat pump as supplementary heat source. The solar hot water volume provided reaches 15.5m^3/d, accounting for 79% of the total hot water

Spread page: Ground water system.

View of Landscape Water

Water-saving devices, sanitary wares and relevant hardware accessories are adopted conforming to National Standard GJ164-2002. Sprinkler irrigation is adopted for watering plants to save water.

Efficient Use of Materials

The project features a simple and elegant architectural style, with the cost of its decorative elements accounting for only 3.5‰ of the total project cost. Recyclable materials are also widely used, constituting 12.8% of the total weight of building materials.

Indoor Environment Quality

Skylights are installed on the rooftops of the auditorium and the media center, so as to improve the daylight conditions of the lounges, conference rooms and foyer, as well as to enhance the building's overall lighting effects. Simulation tests show that the daylight factors of 75.81% of overall functional spaces conform to relevant requirements as specified in National Standard GB 50033—2001.

Air vents are evenly distributed around the building's facades, with more on indoor atriums to facilitate indoor ventilation. Under the boundary conditions of mean wind speed of prevailing wind direction in winter and transitional seasons, the Air Exchange Efficiency of project'ss main function spaces averages two times and above per hour.

Operational Management

Intelligent systems are optimally deployed, with a well-developed information network system.

Collaborations

Kunshan City Construction Investment and Development Co., Ltd. commissioned China Academy of Building Research Shanghai Institute to offer consultancy on green building practices. From the beginning of the design stage, the company convened numerous meetings to communicate with designers from the China Architecture Design & Research Group regarding reasonable green design schemes. During the construction process, the company also had personnel to visit construction site numerous times and communicate with the builders, providing technical support. The good cooperation of all concerned parties contributes solidly to gain the certification of two-star green building.

—

Text: Shao Wenxi, Shao Yi and Han Jingxing, China Academy of Building Research Shanghai Institute

结构设计 Structure design

为实现建筑飘逸灵动的外墙曲面，需要合理确定结构的平面布置及竖向的传力关系，并准确进行模拟，对结构设计至关重要

TO REALIZE THE ELEGANT AND FLEXIBLE CURVED SURFACE OF THE OUTER WALL OF A BUILDING, IT NEEDS TO REASONABLY CONFIRM THE PLANAR ARRANGEMENT OF THE STRUCTURE AND THE VERTICAL POWER TRANSMITTING RELATIONSHIP AND CAN ACCURATELY SIMULATE,WHICH IS CRUCIAL TO STRUCTURAL DESIGN

昆山市文化艺术中心建筑造型源自于昆曲中流动、柔美的水袖，以其造型精美，外立面曲线丰富而著称。项目由大剧院和影视中心及配套车库等组成，是集文化交流、会议展览、休闲娱乐等多项功能于一体的综合性文化建筑。

为实现建筑飘逸灵动的外墙曲面，并对里出外进的曲线进行找型，常需要通过作图法确定结构平立面的相互关系，这使得合理确定结构的平面布置及竖向的传力关系并准确进行模拟，对结构设计至关重要。

大剧院建筑平面大致呈圆形，外轮廓为不规则曲线，且每层各不相同。建筑功能包括1400座的大剧院和位于三、四层会议中心等。结合内部多样的使用功能，并对结构复杂程度做“减法”，大剧院采用框架-剪力墙结构，框架柱布置北段遵循横平竖直的布置方式，南段基本遵循环形方式；剪力墙布置于舞台和观众厅周边，以及楼电梯间。为同时满足曲线外立面的要求，出现了大量的悬挑结构，很多部位的悬挑长度甚至超出常规尺寸。结构设计过程中，为同时满足建筑物内部使用和外部造型的要求，与建筑师进行沟通，对外轮廓曲线的进行了反复多次的调整，以使悬挑长度进行适当的减小；对无法减小悬挑长度的部位，去除部分悬挑端部楼板，以减轻重量。同时，在悬挑长度较大的梁中布设预应力钢筋，保证结构的正常使用。

为满足建筑师对剧场入口大空间的要求，大剧院南侧主入口处，柱子数量不多，且与内部观众厅剪力墙筒体间的楼板多处大开洞区域，连接十分薄弱。结构设计中，对弱连接部位的梁及楼板进行性能化设计；屋面层每根框架柱均设置梁与内部剪力墙筒连接，以保证结构整体性。结构计算采用多种计算模型进行包络设计，并补充单榀验算以保证结构的强度。

观众厅共四层看台。其中一层、二层池座为连成整体。三层楼座为大悬挑看台，且悬挑长度大于压重长度。为减小悬挑看台自重，保证结构安全性，楼座看台采用钢结构。

大剧院观众厅屋面为大跨结构，采用实腹钢梁形式。台口部位结合舞台设备布置，采用钢桁架。主舞台顶呈矩形，尺寸为34m×24m，出主屋面后，为满足设备机房布设要求，顶部呈椭圆形，最大悬挑长度约16m，采用实腹钢梁形式，栅顶层吊挂与屋面层钢梁下。主舞台四角的框架柱采用型钢混凝土柱，并对其进行性能设计，使其受力满足中震弹性设计要求。框架柱周边均布设有剪力墙，计算时，在柱子与剪力墙之间开计算缝，确保框架柱的安全。

观众厅屋面飘带：观众厅屋面的飘带造型为模拟昆曲中流动、柔美的水袖，造型十分复杂，悬挑长度均在25m左右，采用管桁架形式。东北角大悬挑（海角）：悬挑长度达到17m，采用一层通高的钢桁架。

影视中心采用框架结构，为满足建筑造型要求，影厅东侧采用弧形框架柱，并采用中震弹性的性能化目标。屋面造型采用钢结构。

昆山文化艺术中心工程存在结构空旷、楼板不连续、楼板弱连接、结构扭转效应较大、位移比较大等不规则类型，为具有多项不规则的多层建筑。为满足结构的抗震性能设计要求，对结构采用不同的性能化目标，在结构关键部位采用型钢混凝土构件。结构设计过程中，院内进行了多次结构专项评审工作，在江苏省审图中心及有关各方的支持和帮助下，顺利完成施工图设计。

2012年12月26日，昆山文化艺术工程完成竣工验收。在短短的两年半之内，将一座造型精美的建筑屹立在昆山的地面上，在刚刚结束的苏州十大公共建筑评选中名列第四。

—

文：王奇、张根俞、朱炳寅 中国建筑设计研究院结构四室

对页：影视中心屋面大悬挑钢结构。
本页：大剧院四层的超大悬挑（海角）。
Opposite page: Large cantilever steel structure on ceiling of Media center
This page: Large cantilever streel structure on of the theater.

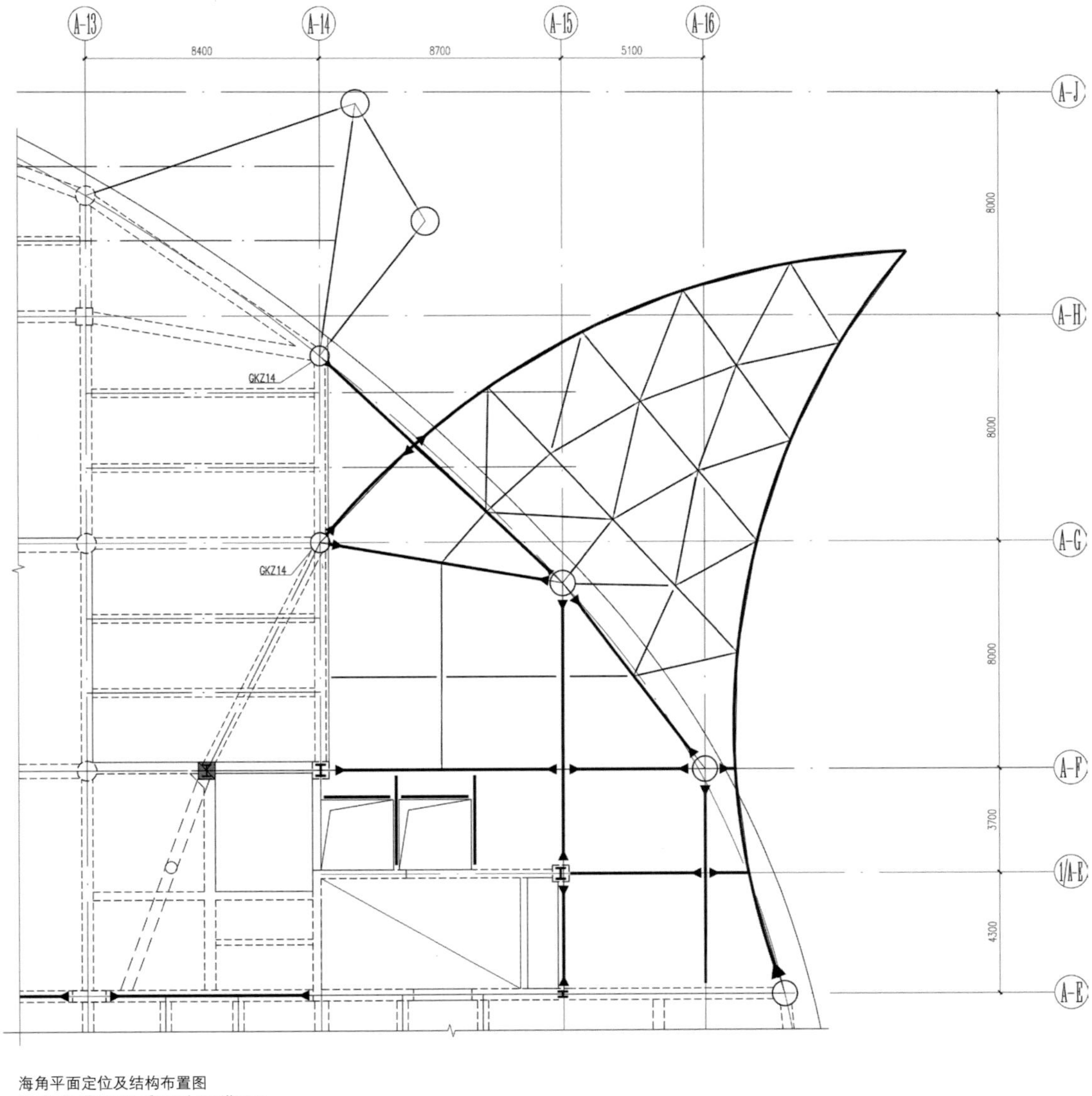

海角平面定位及结构布置图
Analysis diagram of steel cantilever

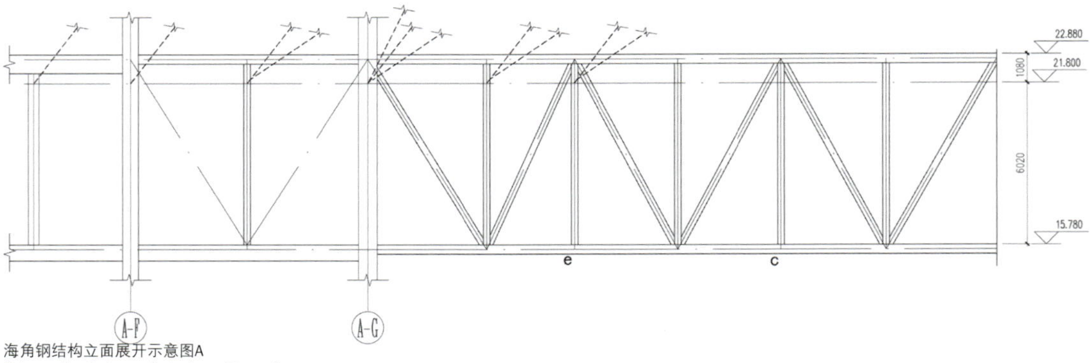

海角钢结构立面展开示意图A
Schematic diagram of steel cantilever A

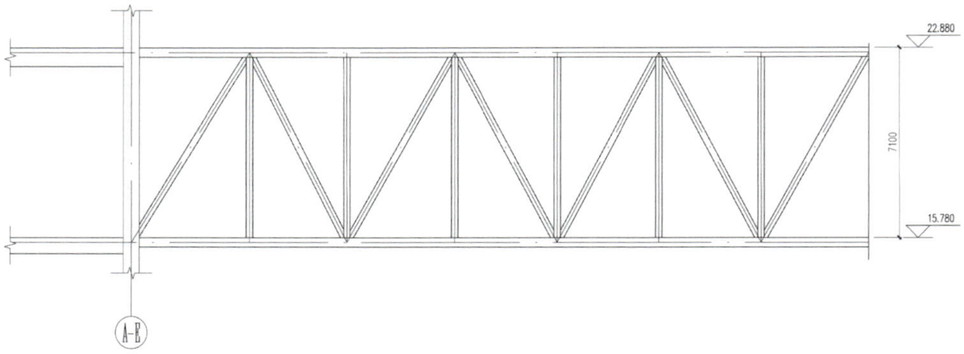

海角钢结构立面展开示意图B
Schematic diagram of steel cantilever B

The building model of Kunshan Culture and Art Center is derived from the flowing and flexible ink water sleeve in Kunqu opera, and it is famous because its beautiful model and abundant curves of the outer vertical surface. The project comprises the theater, the film and television center, matched garages, etc., and it is the comprehensive cultural building which combines culture communication, meeting exhibition, entertainment and other functions into a whole.

To realize the elegant and flexible curved surface of the outer wall of the building, and model the curves which go in and out, generally, it needs to confirm the mutual relationship of the planar surfaces and vertical surfaces through the graphing method, and can reasonably confirm the planar arrangement of the structure and the vertical power transmitting relationship and can accurately simulate, which is crucial to structural design.

Generally, the plane of the theater is round, and the outer contour comprises irregular curves which are varied by floors. The building function includes the theater which owns 1400 seats and meeting centers at the 3rd and 4th floors. Combined with the interior multiple utilization functions, and simplifying the structural complexity, the theater adopts the framework-shearing wall structure. Framework posts are arranged in the north section according to the layout manner of regular transverse and vertical lines as well as the annular manner in the south section. Shearing walls are arranged around the stage and the auditorium and the elevator room. To simultaneously satisfy the requirement of the outer vertical surface of curves, there are a lot of cantilever structures, and the cantilever lengths of many parts are even longer than the normal size. In structural design, to simultaneously satisfy the requirements of the interior utilization and the exterior model of the building, it has been communicated with the architect, and the curves of the outer contour are regulated for a plurality of times so as to prope rly minimize the cantilever length. To the part of which the cantilever length cannot be reduced, and some floor slabs of the cantilever end are removed so as to reduce the weight. Meanwhile, prestressed bars are arranged in the beam of which the cantilever length is longer so as to ensure the normal use of the structure.

To satisfy the requirement of the architects on the big space of the theater entrance, there are not many posts at the main entrance of the south side of the theater, and many big opening areas on the floor slab among the cylinders of the shearing wall of the interior auditorium, and the connection is very weak. In structural design, performance-based design is carried out to the beam and floor slab at the parts of weak connection. Beams are arranged in every framework post of the roofing layer to be connected with the interior shearing wall cylinder so as to ensure structural integrity. The structural calculation adopts multiple calculation models to carry out enveloping design, and replenish single cross checking to ensure structural strength.

There are four floors of stands in the auditorium, wherein the first floor and the second floor are connected into a whole, the balcony at the third floor is the big cantilever stand and the cantilever length is longer than the ballasting length. To reduce the self weight of the cantilever stand, and ensure structural safety, the balcony stand is in the steel structure.

The roof of the auditorium of the theater is in a long-span structure, and adopts the full-web steel beam way. The proscenium opening is arranged by combining with stage equipment, and steel truss is used. The top of the main stage is rectangular, and the size is 34m × 24m. After being out of the main roof, to satisfy the arrangement requirement of equipment machine rooms, the top is oval, and the longest cantilever length is about 16m. The full-web steel beam way is used, and the paling top layer is hoisted under the steel beam of the roof layer. Framework posts at the four corners of the main stage are made of steel reinforced concrete, and performance-based design is carried out to them to make their stress satisfy the design requirement of elasticity level under medium earthquake. Shearing walls are uniformly arranged at the periphery of the framework posts. Calculation fractures are made between the posts and the shearing wall to ensure the safety of framework posts.

Roof ribbons of the auditorium: the roof ribbon model of the auditorium simulates the flowing and flexible ink water sleeve in Kunqu opera, and is very complicated. The cantilever length is about 25m, and the pipe truss way is used.

The big cantilever in the northeast corner (haijiao): the cantilever length is 17m, and the one-floor high steel truss is used.

The framework structure is applied to the media center. To satisfy the requirements of the building model, arc-shaped framework posts are applied to the east side of the movie house, and the performance-based target of the elasticity level under medium earthquake. The roof model adopts the steel structure.

There are irregular types such as spacious structure, un-continued floor slabs, weak slab connection, high structural twisting effect, big displacement, etc., in the project of Kunshan Culture and Art Center which is the multi-irregular multilayer building. To satisfy the design requirement of anti-seismic performance of the structure, different performance-based solutions are applied to the structure, and section steel concrete members are applied to key structural parts. In structural design, a plurality of special structural reviews are carried out in the institute, and with the support and help of Jiangsu Drawing Check center and other relevant parties, the design of the construction drawing is successfully completed.

On December 26, 2012, Kunshan Culture and Art Center was completed and approved. Within the short 2.5 years, the building of which the model is beautiful was established on the ground of Kunshan. The project is ranked at the 4th position in the appraisal and selection for the Top 10 public buildings in Suzhou, which is only over for a few days.

—

Text: Wang qi, Zhang Genyu, ZhuBingyin, NO.4 Structural Design & Research Institute China Architecture Design & Research Group

对页：大剧院外立面。
本页：飘带细部。
Opposite page: Facade of theatre.
This page: Detail of ribbon structure curtain wall

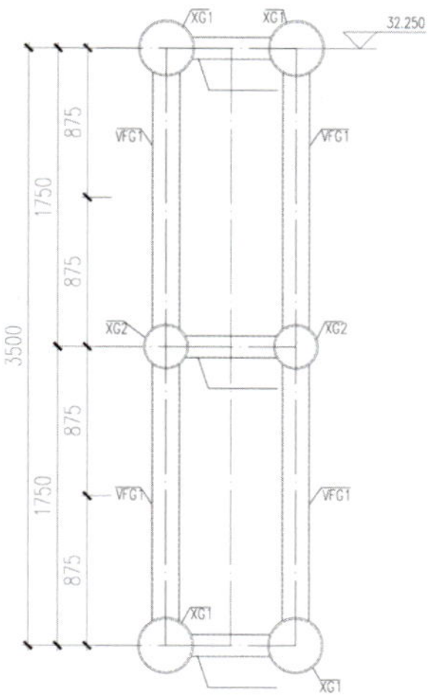

飘带桁架布置示意图
Conceptual diagram of ribbon structure of curtain wall

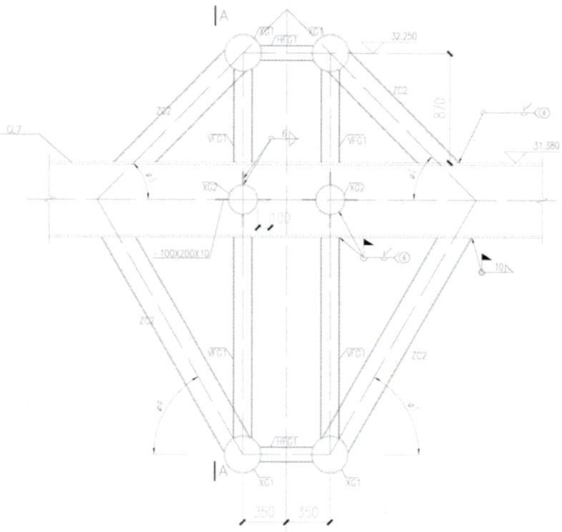

飘带结构与舞台塔桁架连接示意图
Conceptual diagram of connection between ribbon structure truss and stage tower

声学设计

Acoustic Design

声学设计师在设计阶段就参与工作，确保了从声学角度配合建筑师不断完善建筑设计方案，并通过利用计算机软件模拟，获得良好的室内音质

THIS ACOUSTIC DESIGNER TOOK PART IN THE DESIGN PHASE OF THE KUNSHAN CULTURE AND ART CENTER, OFFERING ACOUSTIC ASSISTANCE TO ARCHITECTS IN CONTINUOUSLY IMPROVING THE ARCHITECTURAL DESIGN SCHEME FOR THE STRUCTURE, AND THROUGH THE USE OF ACOUSTICAL SIMULATION SOFTWARE, HAS ACHIEVED GOOD ROOM ACOUSTICS

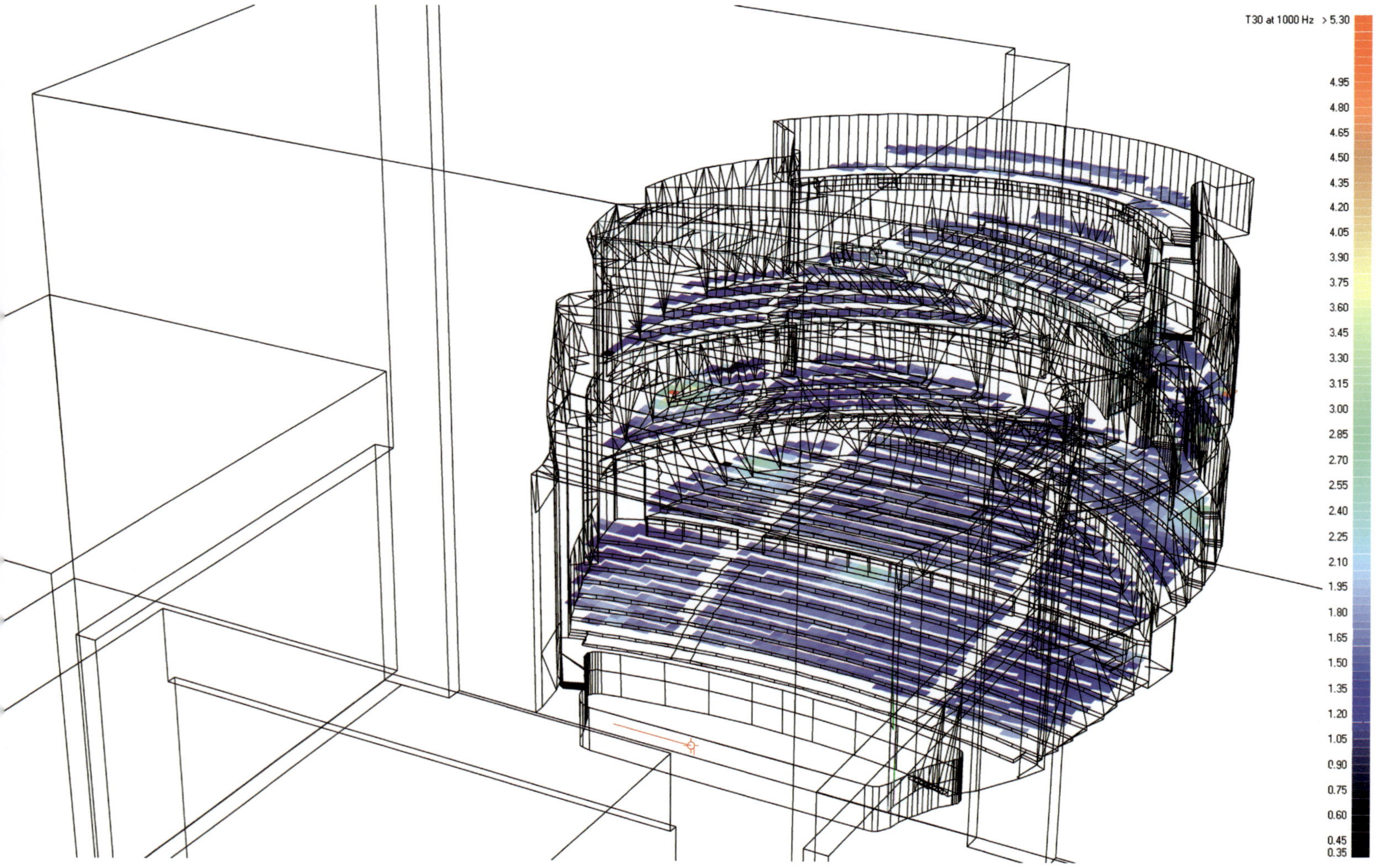

观众厅室内混响时间分析图（500Hz）
Analysis diagram of reverberation in auditorum (500 hz)

昆山文化艺术中心是昆山市最重要的文化建筑，要满足举办各种文艺演出的需求，包括歌舞、戏曲、交响乐、话剧等，给人专业的视听享受，良好的音质是文化中心成功的必要条件。

室内音质包括多个方面，从主观听音角度，良好的音质具有如下特点：声音清晰、同时有良好的丰满度，两者达到理想的平衡；声音亲切自然，感觉演员离你很近；响度合适，听起来很轻松，既不费劲，又不太响；有良好的空间感，让观众感到自己被音乐围绕；没有回声、声聚焦等音质缺陷；没有噪声干扰，很小的声音都能听得清楚；舞台上演员感觉有很好的响应，演唱、演奏感觉良好。昆山文化艺术中心音质设计以上述为目标，力求做到尽善尽美。

有幸参与昆山文化艺术中心的建设是十分幸运的事，我们在2009年文化艺术中心建筑设计之初就参与工作。开始与建设单位、建筑设计人员交流，确定文化艺术中心建设具体目标。从声学角度配合建筑师不断完善建筑设计方案，确保有合理的观众席布置、建筑体形，为获得良好室内音质创造基础条件。在建筑方案确定后，通过计算机模拟，初步判断室内音质效果，检验室内声场是否存在缺陷，协助建筑师进行视线分析，利用计算机软件模拟分析，检验是否有视线及声音遮挡。

合理布置噪声、振动较大的设备，确定适当的空调送风及管道消声方式，确定机电设备隔振及设备机房降噪方案，选择隔声性能良好的墙体及楼面做法，确保文化艺术中心使用中没有噪声干扰；装修设计阶段，与室内设计师沟通协调，确定墙面、吊顶、地面做法，保证其声学性能。

文化艺术中心大剧院观众厅造型新颖独特，为确保声学效果，在室内装修方案基本确定后，我们制作了比例为十分之一的声学缩尺模型，通过在模型中进行声学测试，预先掌握大剧院室内音质。

为获得准确声学数据，同时也方便观察观众厅装修效果，声学缩尺模型制作十分精美，是我们所知国内最好的剧场观众厅声学缩尺模型。施工中，参与材料选择、吸声结构构造做法、设备选型等，保证声学措施得到贯彻落实。观众厅座椅是最大吸声体，占观众厅吸声总量的60%以上，为确保座椅吸声满足要求，对座椅生产提出了具体声学要求，并对座椅样品进行了声学实测。文化艺术中心设计及施工过程中经常需要沟通、协调，该中心从设计到施工，历时4年，我们到现场技术服务超过40次。文化中心大剧院在施工过程中，对大剧院声学进行了三次现场过程测试，并进行了完善。文化中心施工完成后，对大剧院、小剧场等又进行了系统的声学测试，测量结果表明，大剧院、小剧场各项声学指标完全符合剧场设计规范。

经多次演出使用，观众及演员对音质效果反映很好。为使艺术中心大剧院能举办专业交响音乐会，我们在舞台上配置了活动声反射罩。

建成后的大剧院室内观众厅前区墙面及吊顶的角度都是按声学要求严格设计的，可以把早期声音反射到观众厅视线条件好的前区及中区，观众厅后区由中部吊顶及墙面提供早期反射声。

为配合建筑获得完整统一的观众厅室内效果，观众厅墙面及吊顶均采用GRG板，观众厅吊顶及侧墙为高反射GRG板，后墙利用开洞的GRG板做成全频强吸声结构，以控制长延时反射声。

—

文：张三明 浙江大学建筑技术研究所

本页：舞台墙面两种穿孔率吸声结构，面板为8mm厚穿孔硅钙板，内衬无纺布及环保型吸声材料。

This page: There are two types of sound absorption structure for the walls on stage, with different perforated percentages, the panel is 8 mm thick perforated calcium silicon plate, lined with non-woven fabric and environmentally sound absorption materials.

Planned as an iconic cultural landmark in Kunshan City, Kunshan Culture and Art Center (KCAC) is expected to be able to hold all kinds of artistic performances, including dance, opera, symphony and drama, and offer professional audio-visual enjoyment. Therefore, good sound quality is vital for the success of the center.

Room acoustics is comprised of numerous elements. From a subjective perspective, good sound quality shall boast the following characteristics: an ideal balance of clarity and fullness; sounding pleasant and natural, as if the actor is performing right in front of you; an appropriate loudness level, allowing for ease of listening without sounding too loud; a good sense of space, allowing the audiences to feel surrounded by music; no such sound quality defects as echoes or sounding focusing; no noise interference, so that even the slightest sound is audible to audiences; actors feel good acoustically when performing onstage. The acoustic design efforts for KCAC aim to fulfill the abovementioned goals and achieve acoustic perfection.

We are fortunate to be able to participate in the KCAC project from the very beginning. Since we got involved in the projects design stage in 2009, we had extensively communicated with the builders and architectural designers to determine the concrete goals the project aims for, and offered acoustic assistance to architects to continuously improve the architectural design scheme and ensure reasonable layout and architectural form of the auditorium, so as to pave the way for achieving good room acoustics. And after the architecture scheme was determined, through the use of acoustical simulation software, we have also assisted architects in conducting line-of-sight analysis to determine the initial effects of room acoustics and check if there are defects with indoor sound field, or whether the line of sight and sound is blocked.

We also rationally arranged those noisy and vibration-prone equipment, and determined the proper method of air supply of air-conditioners and appropriate duct silencing measures, selected the best vibration-proof plan for mechanical and electrical equipment and optimal noise-reducing solutions for equipment room, and chose the best practices for building walls and floors to ensure good sound insulation capability, so that the center will be effectively protected against noise interference during use; during the phase of decoration design, we actively communicated with interior designers to determine the best practices for decorating walls, ceilings and floors, in an effort to ensure their acoustic performance.

The auditorium of the center's grand theater was very uniquely fashioned. To ensure optimal acoustic effect, after the interior decoration plan was

本页及对页上图：观众厅后部GRG开孔板吸声结构，顶部为GRG板吊顶，具有非常好的反射性能。
对页下图，从上至下：文化艺术中心大剧院室内声学缩尺模型。大剧院座椅声学测试实验室现场。
This page and opposite page, above: Perforated GRG panels are used for the rear wall, high-reflection GRG boards are used to build walls and the ceiling of the auditorium.
Opposite page, below, from above to below: Acoustic scale model of auditorium. Acoustic testing lab for auditorium seating

determined, we created a 1/10 scale acoustic model, and by conducting tests in the model we managed to gain an initial insight into the auditorium's interior sound quality. In order to acquire the most accurate acoustic data, and also make it convenient to observe the decoration effects of the auditorium, we produced a very magnificent scale model, which to our knowledge is also the best acoustic scale model ever created for an auditorium in China.

During the construction process, we also participated in selecting materials, determining best practices for fabricating sound-absorbing structures and selecting appropriate equipment, in an effort to ensure all relevant acoustic measures are sufficiently implemented; in consideration of the fact that auditorium seat is the largest sound absorber, collectively accounting for over 60% of the auditorium's total sound absorption, to ensure that the seats meet relevant sound-absorption standards, we imposed specific acoustic requirements and also conducted acoustic testing on the seating samples. Realizing that the design and construction of the project require constant communication and coordination, during the 4-year design-and-construction process, we visited the site more than 40 times to offer on-site technical services. During the construction process, we also conducted three on-site acoustic tests in the grand theater, with relevant improvements made. After the completion of the construction process, we again performed systematic acoustic testing in the grand theater and the auditorium, with the results showing that all acoustic indicators comply with relevant design specifications. After the facility was put into use, we received positive feedback from the audiences and actors regarding the facility's acoustic performance.

To make the auditorium capable of holding professional symphonic concerts, we installed a movable sound reflector on stage.

The angles of the wall and ceiling in the front section of the auditorium are designed in strict accordance with relevant acoustic requirements, capable of reflecting sound to the front section and middle section with good sightlines, with sound also reflected to the rear section via the middle ceiling and wall.

In order to achieve uniform acoustic effects inside the auditorium, high-reflection GRG boards were used to make the wall and ceiling of the auditorium. Perforated GRG panels were also used for the rear wall to create a structure with a strong capability to absorb full-frequency sound, so as to control the long-delayed reflected sound.

—

Text: Zhang Sanming, Architectural Technology and Research Institute, Zhejiang University

舞台设计

Stage Design

舞台设计与机械设备的配置可以满足各种大型歌剧、舞剧、大型交响乐和大型综合文艺演出的要求，具备了接待国内外各种演出团体的条件

THE CONFIGURATION OF STAGE DESIGN AND MECHANICAL EQUIPMENT MEET THE REQUIREMENT OF A VARIETY OF LARGE-SCALE OPERAS, DANCE DRAMAS, SYMPHONIC MUSIC AND INTEGRATED ARTISTIC PERFORMANCE, WITH THE CONDITIONS TO RECEIVE VARIOUS DOMESTIC AND INTERNATIONAL PERFORMANCE GROUPS

整体设计思路

昆山市文化艺术中心要建成国内一流的文化交流场所，舞台作为整个文化艺术中心的核心。昆山市文化艺术中心是昆山市重点工程，整个工程无论是质量还是技术，都有极高的要求。其最重要的舞台设备工程的设计成果将直接影响整个文化艺术中心建设的档次水平。我们的设计工作是为业主提供一套工艺先进、功能强大的系统方案及高品质的舞台设备和设施。并最大限度保证舞台设备的可用性以及低运作成本和低维护费用。

大剧院

舞台机械设备的配置均能满足各种大型歌剧、舞剧、大型交响乐和大型综合文艺演出的要求，同时兼顾其他类型的演出。做到工艺设计科学、技术性能优良、配置优化实用、运行安全可靠、维修操作方便、经济合理。充分考虑与当时已建成的国内外各大型剧场的机械舞台的功能相同或优于后者，以利于国内外演出的交流。其舞台机械、电气设备的核心关键设备均为行业内国际顶级产品，从根本上确保了舞台机械系统的高品质和高可靠性。

大剧院采用镜框式箱形舞台结构，由主舞台、两侧侧舞台和后舞台组成“品”字形舞台。台下设备配置了主升降台、侧车台以及后车载转台等，可以基本满足演出要求：在主舞台区域设置了六块双层主升降台，用以应对立体造型需要；侧车台及后车载转台的运用可满足演员或多组景物快速进出舞台的输送要求。

为了方便满足布景垂直运输需要，在大剧院主舞台区域后侧设置了一台运景升降台。根据剧院音乐演出的需要，大剧院还设置了一套声反射罩。

大剧院台上设备所有电动吊杆均采用变频调速控制，可吊挂750kg载荷；每道灯杆分为三段，总承载达到2100kg；又配有两侧灯光吊笼，完全满足舞台演出对挂景和挂灯的要求。总的来说，从硬件设施上，剧院已完全具备了接待国内外各种演出团体的条件。

大剧院舞台台下机械分别配有乐池升降护栏1台、乐池升降台2台、主舞台升降台6台、侧辅助升降台12台、侧台车台12台、侧台车台补偿台12台、运景升降台1台、后车载转台1套、后舞台补偿台6台、演员升降小车2台、两会座椅车台2台、乐池座椅台车12台、卸货平台1台等，以及台下机械电气与控制系统。

其中，乐池升降台分为两块，每块都具有5个预停点。乐池升降台利用不同的高度变化，形成各种使用形式。乐池升降台上升至舞台台面高度时，形成舞台的前伸扩展部分；停在观众席前排地面高度时，用于增加观众席前区座位；下降至一定高度时，可以形成的乐池。当乐池升降台下降到低于观众席平面前，乐池升降护栏升起，防止人或物掉入乐池内，保证演员和观众的安全。

在主舞台区域设置的6台双层主升降台（3mx18m）是现代化机械舞台的主体，也是该剧院台下舞台机械设备最重要的组成部分，它能够灵活、丰富地变换舞台形式，使整个主舞台在平面、阶梯之间变化。通过升降台相互组合，改变升降高度，可形成不同的演出平台；与车台组合使用，可用于各种歌剧、舞剧和综艺演出，搭装场景，使布景在演出中多次快速迁换，参与演出，以增加表演效果。升降台还可以降到舞台面下4.5m，用于运送演员或道具。

在两侧侧舞台各配置6台侧台车台（3mx18m）。当主升降台下降至车台高度后，车台可以开到主升降台上，并可随主升降台升降，既可以迁换布景，也可以参与演出。同时侧台车台补偿台可上升补平舞台面，保证人员安全。

后舞台配置了1台车载转台（18mx18m），车载转台上配置了一个直径为12m的内转台和一个直径为17m的外转台（转环）。在主舞台区域后侧设有一套台运景升降台，运输升降机的台板能够停留在行程范围的任何位置，满足布景的垂直运输需要。

剧场配置了演员升降小车，用于运送道具和演员，满足舞台特殊场景演出和特技演出。另外配置了“两会”用座椅车台，用于大型会议使用。使用时每台座椅车台放在三个双层主升降台上，不用时储存在后台下方的储存室。剧院货物进出口设置了可以电动调节高度的卸货平台，用于不同运景车辆的货物装卸。

大剧院舞台台上分别配置了台唇（台口外）单点吊机、台唇链式吊机、台唇吊杆、防火幕、台口吊杆、大幕机、假台口、侧布景吊杆、电动吊杆、灯光吊杆、二道幕机、灯光吊笼、主舞台单点吊机、天幕吊杆、舞台吊杆、声反射罩、侧台装景吊机等机械设备以及台上机械电气与控制系统。

其中，用于为乐池上方提供灯光或布景吊点的8台唇单点吊机，也可用于吊装台前饰物、电子显示屏和音响等设备。用于吊挂线阵列扬声器的4台唇链式吊机；台唇吊杆用于吊挂台口横向LED字幕屏。台口处设置了1道安全防火专用的刚性防火幕。设置于舞台台口内侧的假台口，由侧片和上片组成。通过侧片和上片位置的变化，可以改变舞台开口大小。大幕是用来分隔舞台与观众厅的第一道幕，具有对开、提升及斜拉等功能，是使用最频繁的舞台机械。2套二道幕机是分隔舞台表演区的专用设备，具有安装简单，更换灵活等特点。二道幕机可根据演出需要吊挂在任意一道电动吊杆上，具有对开、升降两种功能。

设置于主舞台上部的3道台口吊杆和54道电动吊杆，用于提升布景、各种幕布、二道幕和吊挂灯具等。电动吊杆采用变频调速控制，最高运行速度1.5m/s，可根据演出需要，几台组合或同步运行。主舞台上空两侧设置了2道侧布景吊杆，用于吊挂侧幕或侧景片。主舞台上部配置了4道分段式灯光吊杆和12台灯光吊笼以充分满足主舞台区演出的灯光要求。主舞台上空后侧设置了1道天幕吊杆用于吊挂天幕。12台主舞台单点吊机可以在主舞台上空任意需要位置吊挂景物或演员，方便演出特技，可单机起吊，也可多台组合使用。剧场还设置了一套声反射罩供剧场音乐演出的需要。两侧舞台上空各配置了4台装景吊机，使侧舞台装运景片更方便快捷。后舞台设置了15道电动吊杆，用于舞台演出使用。

多功能剧场

多功能剧场主要用于昆曲演出，剧场形式采用开敞式舞台布局，是根据最灵活的使用方式来设计，配备了移动座椅以及平整地面上的灵活舞台。

多功能剧场中均可以实现以下多种舞台布局方式：

·镜框式舞台：座椅面对舞台，可以使用幕布表示台口

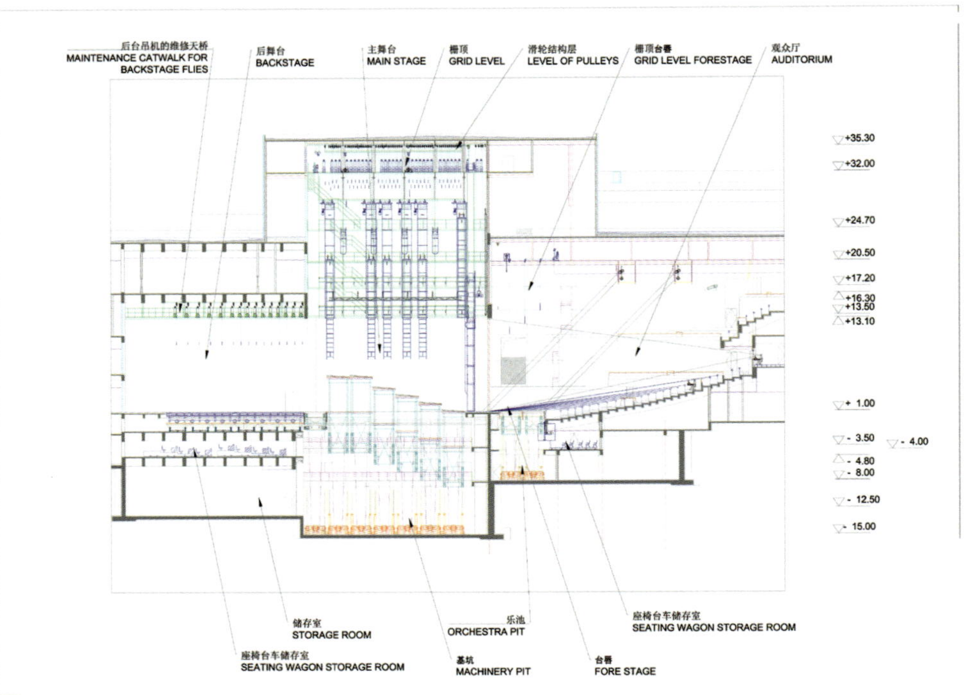

大剧场俯视剖面布置图
Cross section of the theatre

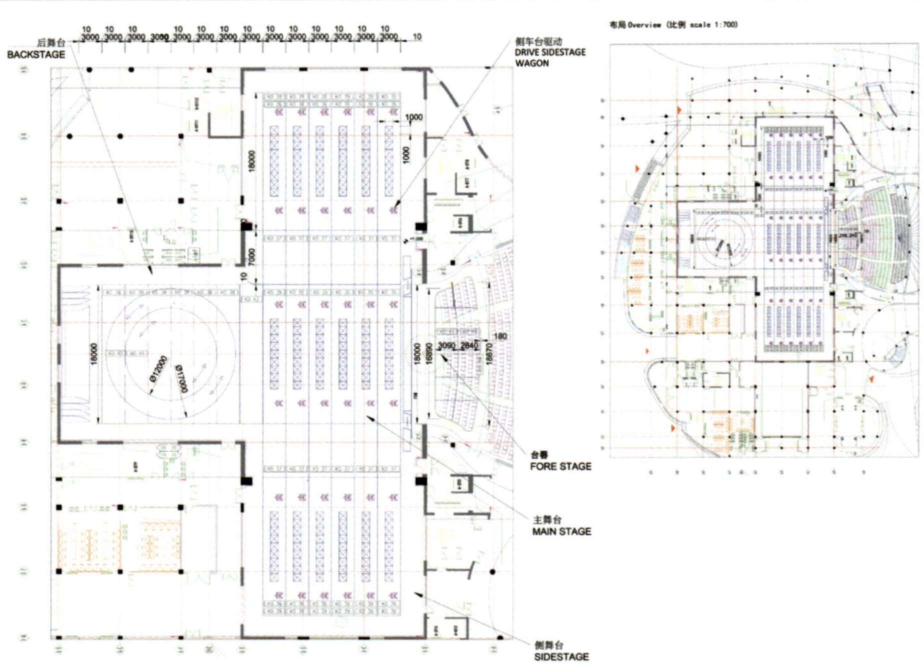

大剧场台下布置图
Layout drawing of under the stage of theatre

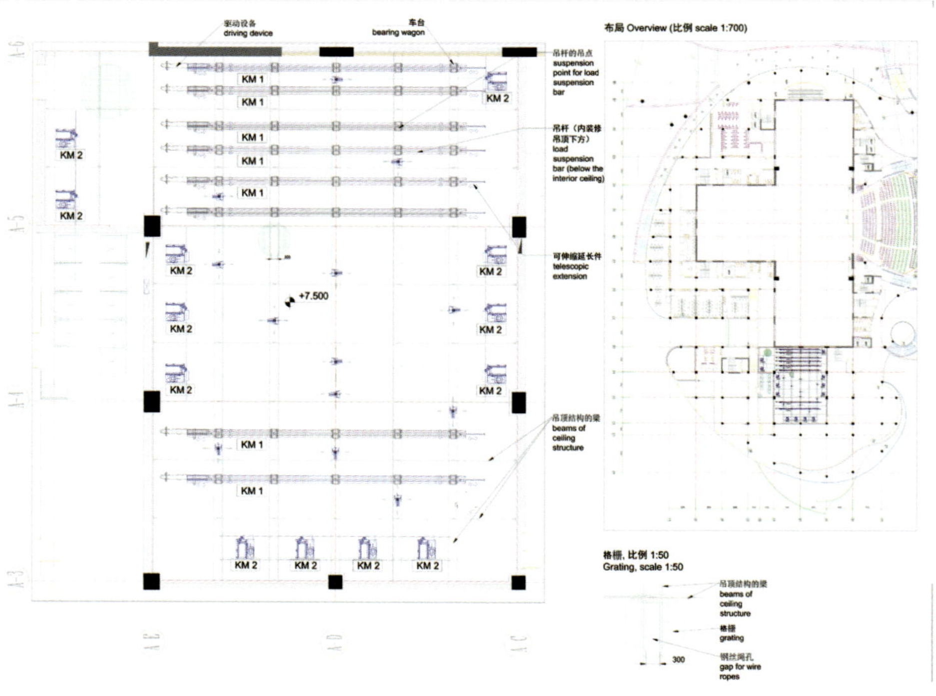

小剧场（多功能剧场）台上布置图
Layout drawing of on the stage of the small theatre (Multi-function Hall)

（例如昆曲演出、其它小型演出、报告会、会议等）

·开敞式舞台：座椅面对延伸舞台，无台口（例如表演、更大型的报告会、艺术展出等）

·T台：舞台在中间延伸，两面环绕座椅（例如演出、报告会活动、公司会议等）

·延伸的T台：与T台相同，中间部分延长一倍（例如时装秀、产品发布等）

·岛式舞台：舞台四周都是座椅（例如介绍会、戏剧、特殊演出等）

·宴会厅：只有宴会座椅（桌子，椅子），无舞台（例如宴会活动、鸡尾酒会）

为了实现这些布局，多功能剧场台下设备配置了21块升降台，这些升降台配备机电驱动系统，可以安全、快速、方便地转换布局。为使观众厅座椅灵活，提供了18台座椅台车，可以根据每场演出的需要来调整这些座椅排列布局。如果不需要用座椅台车，座椅台车可以储存在升降台的下层台板上。

台上机械由8道电动吊杆及13台单点吊机组成，其均采用变频调速控制，其中每道吊杆可吊挂500kg载荷，每台单点吊机可吊挂250kg载荷，用于幕布、装饰物以及厅内灯光和音响设备的悬吊。电动吊杆沿整个厅纵轴方向分布，能够水平移动，因此电动吊杆位置可以根据每场演出要求灵活选择。单点吊机卷扬机将固定安装，每台卷扬机将配备一组移动滑轮，用于悬吊钢丝绳，这样几乎可以在整个多功能剧场内实现吊点的布置。

厅内设计了一道对开大幕，但由于舞台和观众厅位置灵活多变，此大幕能够通过任何一个电动吊杆悬吊。从硬件设施上，本剧场已完全具备了接待国内外各种演出团体的条件。

控制系统

控制系统采用人工智能化管理，同时也可进行人工干预，并备有完善的安全保护及应急措施；人机界面友好，显示功能直观，故障诊断功能完善，具有自动、手动两种控制功能。

在任何情况下都不会导致操作故障或程序及数据的丢失。并采取适当的EMC(电磁兼容性)措施和提供适当的 UPS (不间断的电源供给)系统。

结语

经过了近4年的设计及施工，完工的昆山文化艺术中心舞台机械系统能够完美的实现最初的设想，比较圆满地满足各类演出要求。

—

文：李跃文 昆克剧院工程咨询（北京）有限公司

前页及对页：大剧院上空，舞台机械台上设备全部为电动设备，用来吊挂幕布，景片及灯光设备等。
本页上图:多功能厅可以实现多种舞台形式，台下设有多个升降台，并配置了相应的座椅台车，台上设备位置可以移动，以便用于各种舞台形式的灯光及布景。
下图:大剧院主舞台区域台口的舞台设备及观众厅室内实景效果局部。

Previous opposite page and Opposite page: The stage hanging light equipment such as lighting suspender and light basket above the back of main stage, and motorized winch battens used to hang on the curtain, stage scenery.
This page, above: Multi-function Hall can complete various kind of stage effect,The lifting platforms are set under stage and equip the seat trolleys are provided, so the implement on the stage can be moved and satisfied the different light and scenery on the stage.
Below: Stage equipments at the entablature in the area of main stage and partial detail of interior of the auditorium.

Overall design concept

Kunshan Culture and Art Center will be built into the domestic first-class cultural exchangeorganization, and the stage is viewed as the core of the center. Kunshan Culture and Art Center is a key project in Kunshan. The requirement is very high in both the quality and technology of the whole project. The most important design result of the stage equipment project will directly affect the level of the construction in the center. Our design is to provide owner a set of system scheme with the advanced technology and powerful function and the high-quality stage equipmentsand facilities to ensure the availability, low operating costs and low maintenance costs of the stage equipments to the maximum extent.

Auditorium

The configuration of stage machinery equipment meets the requirement of a variety of large-scale operas, dance dramas, symphonic music and integrated artistic performance, while taking into account the performance of other types, which is scientific in process design, high in technical function, optimized and practical in configuration, safe and reliable in operation, convenient, economical and reasonable in maintenance. We have considered the function of mechanical stages in the large-scale theatres at home and abroad at that time, which was at the same level or superior to the latter, being beneficial to the exchange of performance at home and abroad. Its key equipments of stage machines and electrical equipments are the international top-level products to ensure the high quality and reliability of stage mechanical system fundamentally.

Coliseum adopts the picture-frame box-like stage structure, which composes "品" stage by main stage, side stages of two sides and back stage. The main lifting platform, side stage wagon and back vehicle-mounted turntable are set under the stage to meet the requirements of performance: in the area of main stage, 6 double-deck main lifting platforms are set to meet the requirements of three-dimensional contouring; the side car turntable and drive turntable can meet the transportation requirements of rapid passing in and out of the stage for the performers and multi-group sceneries.

In order to meet the requirements of setting vertical transportation, a scenery lifting platform is set at the back side of the area in the main stage. According to the requirements of performance in theatre music, a set of acoustic reflection shell is installed in the coliseum.

All the motorized winch battens of the equipments are controlled by frequency control, which can hang load of 750kg; each lamp standard is divided into 3 sections, and the total load bearing reaches 2,100kg; lighting baskets are equipped in both sides to totally meet the requirements of hanging scenery and light in stage performance. In general, as for hardware facilities, the coliseum can totally have the condition to receive all performance groups at home and abroad.

There are many machines equipped with under the stage, such as one promotion and demotion guardrail of orchestra pit, two lifting platforms of orchestra pit, six lifting platforms of the main stage, twelve lifting platforms of side auxiliary, twelve trolleys of side stage, twelve compensation tables of side trolleys, one scenery lifting platform, one drive turntable of back trolley, six compensation tables of back stage, two lifting cars for performers, two seat turntables of dual purpose, twelve seat trolleys of orchestra pit, a landing platform, mechanical and mechanical and electricity and control system.

Lifting platform of orchestra pit is divided into twosections and each section has five previous hold points. Various shapes have been formed by different altitude variations of the lifting platform in the orchestra pit. When it stops on the floor equal to that of the front row of audience, it is used to increase the seats of front area of audience; the orchestra pit cab be formed by descending some height. When the lifting platform drops and is lower than the flat surface of auditorium, the lifting guardrail is rising to prevent people or things drop into the orchestra pit and ensure their safety.

There are six double-deck lifting platforms in the area of main stage (3mx18m), which is the main body of the modern mechanical stage and the most important part of the mechanical equipment under stage; they are flexible and rich enough to change the stage forms, then the main stage changes between the flat surface and step. Through the mutual combination of lifting platform, the lifting height can be changed to form different performance platform; with turntable, it can be adopted in various operas, dance drama and variety show, which makes the setting can rapidly move and change in many times to participate in performance and improve the performance result. The platform can drop to 4.5m below the stage surface to transport performers and stage properties.

There are 6 side stage trolleys in two sides of the stage (3mx18m). When the main lifting platform falls to the height of turntable, the trolleys can drive to the main lifting platform and can change the setting and participate in performance as the main lifting platform goes up and down. Meanwhile, the compensation platform of side stage trolleys can go up to the filling-in stage surface to ensure the personal safety. One vehicle-mounted turntable is installed on the back stage; one inner turntable whose diameter is 12m and one out turntable whose diameter is 17m are configured in the vehicle-mounted turntable (swivel). Behind the back area of main stage, there is one lifting platform to transport the scene; the platen of transportation lifting can stay in any position in the route range to meet the requirements of vertical transportation of setting.

A performer lifting car is set to transport the properties and performers, which is satisfied with the performance of special scene show and special effect. In addition, the double-purpose set turntable is equipped with to use in bull session. When using, each seat turntable is in 3 double-deck main lifting platforms; without using, it is stored in the below warehouse of backstage. At the entrance and exit of goods, there is the landing

本页:大剧院主舞台栅顶层设备布置，整齐的设备排布及整洁的层面布置，使得舞台栅顶层面便于技术维修人员工作及通行。
对页:在大剧院主舞台后部上空的舞台灯光吊杆及灯光吊笼等灯光吊挂设备。
This page: The layout of the equipment on the top of grid ceiling of theatre's main stage is neat and clean, making the top floor convenient to technical maintenance personnel work and pass through.
Opposite page: The stage hanging light equipment such as lighting suspender and light basket above the back of main stage.

platform that can electronically adjust the height, which is used in cargo handling for different scene vehicles.

There are many equipments and machines configured in the stage of coliseum, such as apron stage (outside the entablature) single point crane, apron stage chained crane, apron stage suspender, fireproof curtain, entablature suspender, big curtain machine, side setting suspender, motorized winch batten, lighting suspender, two screen machines, lighting basket, lighting basket of main stage, backdrop suspender, acoustic reflection shell, side stage setting crane, stage mechanical electricity and control system.

And the eight apron stage single point cranes used for providing light or setting lifting points over the orchestra pit can also be used to lift the ornaments before the stage, electronic displays and stereos, etc. Four apron stage chained cranes are used for hanging linear array loudspeakers. The apron stage suspender is used for hanging the lateral LED subtitle screen at the proscenium where a rigid fire-curtain is set specially for safe fire protection. The false proscenium set inside the proscenium of the stage consists of hard teasers and tormentors. Varying the locations of hard teasers and tormentors can change the size of the opening of the stage. The house curtain is the first curtain used to separate the stage from the auditorium and can run from opposite directions, be lifted and be cable-stayed. It is the most frequently used stage machine. Two sets of second curtains are special devices used to divide the acting area of the stage. They are easy to install and flexible to change. The second curtain can be hung on any motorized winch batten as required by the performance. It can run from opposite directions and be lifted.

The three proscenium suspenders and fifty-four motorized winch battens set above the main stage are used to lift sets, various curtains and second curtains and hang lamps. The motorized winch batten adopts frequency control with a maximum running speed of 1.5m/s. Several motorized winch battens can be combined or synchronized according to the needs of performances. At the two sides above the main stage are set two side suspenders for setting used for hanging side curtains or side scenic pieces. Four sectional lighting bars and twelve lighting baskets are configured in the upper part of the main stage, so as to fully meet the lighting requirement of the performances in the main stage area. A backdrop suspender is set at the rear side above the main stage, which is used for hanging backdrops. Twelve single point cranes for the main stage can be used to hang scenes or performers in any required location above the main stage, facilitating the performance of stunts. They can be used individually or collectively. The theater is also equipped with an acoustic reflection shell to meet the needs of musical performances in the theater. Above the stages at both sides four scene loading suspenders are set respectively, making it easier and faster to load scenic pieces on the side stage. Fifteen motorized winch battens are provided for the rear stage, which is used for stage performances.

Multi-function theater

The multi-function theater, which is mainly used for Kunqu operas is provided with an open platform and designed to maximize its flexibility. It is equipped with mobile seats and a flexible stage on the flat ground.

The multi-function theater allows for several stage layouts as follows:

• Proscenium Stage: seats face the stage. The curtain can be used to represent the proscenium (for example, performance of Kunqu operas, other small-scale performances, report meetings and conferences)

• Open stage: seats are faced with the extended stage. There is no proscenium (for example, performances, large-scale report meetings and art exhibitions)

• T stage: the stage is extended in the middle and surrounded by seats on the two sides (for example, performances, report activities and corporate meetings)

• Extended T stage: it is the same with T stage and the length of the middle part is doubled (for example, fashion shows and product release)

• Arena stage: the stage is surrounded by seats (for example, orientation meetings, operas and special performances)

• Banquet hall: there are only seats for banquets and no stages (for example, banquets and cocktail parties)

In order to implement these arrangements, twenty-one lifting platforms are set under stage in the multi-function theater. These lifting platforms are provided with electromechanical drive systems so that their layouts can be safely, rapidly and easily changed. In order to make the seats in the auditorium flexible, eighteen seat trolleys are provided, the layouts of which can be adjusted according to the needs of each performance. If the seat trolleys are unnecessary, they can be stored on the lower platens of the lifting platforms.

The machines on stage consist of eight motorized winch battens and thirteen single point cranes which all adopt frequency control. Each motorized winch batten can hang up to 500kg of load and each single point crane can hang up to 250kg of load, used for hanging curtains, ornaments, lamps and sound equipment in the hall. The motorized winch battens are longitudinally distributed in the hall and can move horizontally, so their locations can be flexibly selected according to the needs of each performance. The hoisters of single point cranes will be fixed and each hoister will be equipped with a set of mobile wheels used for hanging steel wire ropes so that lifting points can be set almost in the whole hall of the small theater.

A house curtain running from opposite directions is designed in the hall. However, as the locations of the stage and auditorium are flexible and changeable, this house curtain can be hung by any motorized winch batten. In terms of hardware facilities, this theater has been fully equipped with conditions for receiving various domestic and foreign performing groups.

Controling system

The control system adopts AI-based management and allows for manual intervention and is provided with complete safety protection and emergency measures. It features a friendly human-computer interface, visual display, complete fault diagnosis and can be controlled automatically and manually.

In any case, it won't cause operation errors or loss of programs and data. It adopts proper EMC (electromagnetic compatibility) measures and is provided with an adequate UPS (uninterrupted power supply) system.

Conclusion

After almost four years of design and construction, the finished stage machine system of Kunshan Culture and Arts Center can perfectly realize its original assumption, meeting the requirements of various performances.

–

Text: Li Yuewen, Kunkel Theatre Engineering & Consulting (Beijing) Co., Ltd.

对页：大剧院主舞台基坑部分，图中显示为主升降台驱动装置及电缆连接设施。

本页，从上至下：大剧院主舞台栅顶隔栅下部设备——侧灯光吊笼及吊杆体及幕布（其中栅顶距舞台面31m）。主舞台吊杆提升至上部的实景效果。

Opposite page: The foundation ditch of the main stage in theatre, photo is the actuating devices and cable connection facilities of main lifting platform.

This page, from above to below: The bottom equipment of grid ceiling and retainer of main stage, photo is showing the lighting baskets equipped in sides, chained crane and the curtain. The real effect of the main stage chained cranes hoisting to the top which equipped under the grid ceiling and retainer of theatre's main stage.

照明设计 Lighting Design

大剧院的灵感来自于碧水、窄桥、捧花，与昆曲中流动、柔美的水袖。照明设计也同样明显或含蓄地突出了这一主题

THE THEATER IS INSPIRED BY UNIQUE LOCAL ELEMENTS AS GREEN WATER, NARROW BRIDGE, FLOWER BOUQUET, AND FLOWING "WATER SLEEVES" OF THE KUNQU OPERA. THE STRUCTURE'S LIGHTING DESIGN HIGHLIGHTS THE THEME, EXPLICITLY AND IMPLICITLY

本页：夜景照明实景。无论白天还是夜晚，建筑始终体现的是素雅的风格。
对页：测试样板。
This page: The scene during night. The building is always presenting a simple but elegant style in both daytime and night .
Opposite page: Testing sample.

大剧院的灵感来自于碧水、窄桥、捧花，与昆曲中流动、柔美的水袖。建筑从飘逸灵动的外墙曲面，到精致的、犹如刺绣的表面图案，到依水而建、空灵而通透的景观设计，乃至如花朵般绽放的室内空间，无不贯穿着这一整套语言。与之相呼应的，灯光也同样或者明显，或者含蓄的凸出了这个主题。

立面

在日间，穿孔铝板遮住建筑内部，形成屏障；而夜间灯光从穿孔铝板形成的细腻图案背后悠然点亮，宛如帷幕拉开，建筑在夜间呈现出优雅、神秘而富有韵味的风貌。如何含蓄而恰到好处的照亮双层穿孔铝板颇具挑战，简单的使用LED点阵不仅会对日间的建筑形体造成影响，而且会造成眩光及缺少进深感。设计师通过许多次模拟后，采用了独具匠心的设计：在双层穿孔铝板之间的空间，利用安装卡件的设计不仅将灯具稳定固定，并且上层的安装卡件将下部的灯光完全封闭在独有的空腔内，在穿孔的空腔内壁形成多次反射，使得访客随着视角的不同而欣赏到穿孔板内壁的呈光效果。
利用LED点光源成为间接光照明，避免眩光的基础上实现有进深有层次的照明感受，在特殊的场景下亦可将戏剧化舞台化的多媒体效果演绎得淋漓尽致。

室内

灯光让室内空间成为融入环境中的风景，内与外的交流、明与暗的互换，为访客提供了一个富有艺术氛围的共享空间。在设计中着重强调了垂直面的照明，也就是通过着力于照亮建筑的室内立面，展示建筑元素本身的细节，同时保持水平面上较低的照度，这样访客可以更加为整个建筑空间所吸引、融入室内环境，达到空间与人自身的和谐。
另一方面，这样的手法也有利于降低能耗，实现可持续的照明。同时，室内的立面在夜间透过大堂的落地窗，温暖的光环境仿佛晶莹剔透的水晶一般呈现在行人眼前，与外立面的照明相呼应，成为夜色中和谐的一部分。
纯白色简洁空间内，从地面到灯光，从桌椅到楼梯，处处都是行云流水般的弧线设计，让整个空间充满流动感。顶棚灯槽与光带沿着顶棚曲线盘旋，与环境天衣无缝的融合在一起，柔和的光照亮了曲面和地面，在凹凸之间，空间变得丰富。而光与影让这种丰富变得流动，有层次而意味深长。光影流转之间，从暖色到冷色的变幻令人觉得空间丰富起来，充分感受到建筑现代洗练的表达，与柔和含蓄的神韵。

景观

在夜间，通过灯光引导视觉与人们的行进路线，重塑了景观空间。景观灯光相对低调，因为身处景观之中，建筑才是视觉焦点。景观灯光只需要和谐的配合建筑，在正确的角度下引导视线，并提供安全性的照明即可。静水中倒映的优雅建筑，本身就是最好的风景。

夜景与深化实施

建筑夜景灯光是以建筑为载体，使建筑白天形象在夜晚得以延续和扩展的一种表现形式。夜景灯光对于建筑形象的价值，通常比建筑白天的形象更具有感染力。因为有了夜晚黑幕的背景，有光的部分就凸显了出来，更容易形成视觉焦点，加深人们的记忆。所以，在现代建筑的形象表现中，夜景灯光日益被重视。
真正建筑夜景照明设计应该是一种逐渐步入灯光环境艺术领域的学科。在昆山文化艺术中心项目夜景灯光的设计中，照明设计师正是立足于这一设计思想，将“夜景照明”提升为“夜景艺术”，充分运用各种现代照明手段及控制技术，创造出极具艺术感染力的建筑夜景景观。
建筑形体的灵魂来自昆曲水袖意象，所以夜景灯光既要表现出建筑基本的出水芙蓉似的静态美，尤其要表现出“水袖”的灵动之美。对于建筑的静态效果，用常规的泛光照明即可实现，比较容易处理。但要表现“水袖”的灵动之美，则需要花费一些心思。为此，来自德国LEXO的照明设计师想到了将智能灯具藏入建筑外表面的金属装饰穿孔幕墙中，通过灯光控制系统控制灯具明暗和颜色变化，从而在幕墙上形成动态图案的想法。LED灯具体积小巧、长寿命、颜色及亮度可控的特性是智能灯具不二选择。通过设计前期幕墙反复小样模拟试验，选择的LED灯具竟然能够正好完全隐藏在金属幕墙里，而其发出的光打亮了邻近的金属孔，可以组合出完美的图案。这一照明设计的领先之处在于幕墙上形成的图案是用间接光组合而成的。相比常规的LED大屏显示（LED光源直视发光），间接光就显得相当柔和，饱满而安静，正好契合昆山文化艺术中心艺术化的氛围。
LED灯具与金属幕墙的结合安装也着实让深化设计的工程实施者花费了很大心血。最后，通过对双层金属幕墙结构深入的研究和现场持续的安装试验，北京良业照明技术有限公司发明了滑轨架子安装法。在金属幕墙里找到了唯一的安装通路，避开幕墙结构中碍事的螺栓，将预先固定好LED灯具的滑轨安装架子巧妙地滑入双层金属幕墙中，使数量庞大的LED灯具安装俨然变成了一种享受。
昆山文化艺术中心LED夜景照明的播放场景是项目成功的关键因素。昆曲水袖、江南水墨、风、荷、雾、水波等场景将依次出现，抽象唯美是场景的主要基调。包括工程实施方、业主、建筑师、照明设计师及提供LED场景编辑与播放系统的灯具供应商等相关人士一起积极沟通协作，精心策划播放场景，使场景的无穷变化直至各方最终满意成为可能。在各方努力下，该项目夜景照明在“第八届中照照明奖”评选中获得了照明设计奖二等奖。

—

资料提供：黎欧思照明（上海）有限公司、北京良业照明技术有限公司

Lighting Design

The auditorium is inspired by such unique local elements as green water, narrow bridge, flower bouquet, and flowing "water sleeves" of the Kunqu Opera. From its exterior such as the elegant curve of walls, the exquisitely "sculpted" and embroidery-like surface, as well as ethereal yet transparent landscape, to its richly decorated interior, this unique language of design was universally "spoken". Accordingly, the structure's lighting design highlights the theme of the project, explicitly and implicitly.

Facade

During the day, the perforated aluminum panels serve to block the sunlight and shield the interior; and at night, lights are glimpsed gracefully streaming out from behind its exquisitely sculpted facade, creating the visual effect of a curtain slowly unfurling, thus lending a tantalizing sense of grace, mystery and intrigue to the building. How to illuminate the dual layers of perforated aluminum panels in a subtle yet fulfilling manner constitutes the major challenge for the structure's lighting design. The mere adoption of LED dot-matrix display might not only compromise the visual contour of the structure during the daytime, but also cause glare and ruin the visual depth of the building. After conducting numerous simulations, lighting designers finally came up with ingenious design approaches to install fixtures in between the two layers of perforated aluminum panels, which, in addition to fastening lamps in place, also has another advantage: the upper fixtures can serve to completely seal the lower lighting within a "cavity", causing the light to undergo numerous reflections across the wall of the cavity, thereby treating visitors to varying visual "feasts" as they have different angles.

To adopt LED point lights as the indirect lighting source can effectively prevent glare and enhance lighting depth, and on special occasions can also enable to showcase dramatic lighting effects of multimedia.

Interior

Lighting helps incorporate the interior space into the overall visual presence of the building. The indoor-outdoor exchange and the light-darkness interplay help inspire an artistic atmosphere in the shared space for visitors. The lighting design specially emphasizes on vertical illumination, i.e. by highlighting the interior walls to showcase the details of architectural elements, while also maintaining a low level surface illumination thus to make visitors get attracted by the whole architectural space and blend in the interior atmosphere, so as to achieve harmony between space and themselves; on the other hand, this kind of approach can also help reduce energy consumption and achieve sustainable lighting. In addition, the lighting system will cast a warm glow onto the interior walls during the night, lending mesmerizing crystal-like sparkle, which along with the facade lighting make for a harmonious nightscape.

Within the pure white, simple space, elegant curves dominate in the design of everything, from the floor to the lights, from the desks and chairs to the staircases, effortlessly instilling a soothing sense of fluidity. Ceiling lighting grooves and lighting belts spiral along with the curves of the ceiling, melting seamlessly into the surroundings, with soft glowing lights illuminating curves and the floor, lending a rich texture to the space. The play of lights and shadows also makes the texture more dynamic, arousing a sense of depth and intrigue. With the contrasting lights and shadows and the interplay of warm and cold hues having greatly enriched the space, visitors are blissfully allowed a full taste of the subtle glamour of a neat, stylish piece of modern architecture.

Landscape

During the night, lighting is used to guide people's route, effectively helping reshape the landscape space. The landscape lighting is relatively dim, for the building itself shall be the focal point amid the landscape. The landscape lighting needs only to complement the building, and guide people to the right direction, while also providing security lighting. The graceful reflection of the building on a tranquil pool of water is a captivating view in itself.

Nightscape and design implementation

Night lighting of buildings is a way to continue the presentation of a structure's view. It is usually more important to a building in that a night-lit building is more visually appealing than its daytime appearance. Thanks to the dark backdrop provided by the night, the illuminated parts of a building tend to be more easily highlighted, with more easily formed focal points, thus making it easier for the building to impress people deeply. For this reason, increasing attention has been paid to night lighting in modern architecture.

A real night lighting design shall be an art in itself. For the night lighting design of Kunshan Culture and Art Center, lighting designers have treated "night lighting" as "nightscape art", by fully exploiting all kinds of modern lighting techniques and controlling technologies to successfully create an artistically appealing nightscape for the building. The Kunshan Culture and Art Center is an architectural masterpiece by the prominent Chinese architect Cui Kai, who drew inspiration from the "water sleeves" of the Kunqu opera. Therefore, the night lighting design of the building not only highlights the static aesthetics of the structure formed like a lotus emerging out of water, but also accentuates the poignant visual dynamics of the building shaped like flowing water sleeves. It is relatively easy to achieve good "static" lighting effects, as regular floodlighting can make that happen. It is the water sleeves part of the lighting design that was truly difficult and took a lot of consideration.

To fulfill the demanding task, lighting designers from LEXO adopted the method of embedding intelligent lamps into the curtain walls of decorative perforated metal on the external surface of the building, through a lighting control system to manage the brightness and color, forming dynamic patterns on the curtain walls. LED lamps are compact and durable, and also adjustable in color and brightness, making it an ideal choice for intelligent lamps. After conducting many simulations, the chosen LED lamps perfectly fit into the metal curtain walls, with the lights they emit illuminating the neighboring metal holes, creating perfect patterns that fulfill our needs.

How to fix the LED lamps onto the metal curtain walls was another challenge for the engineering contractor—Beijing Landsky Lighting Technology Co., Ltd. After conducting in-depth research into the structure of two-layer metal curtain walls and performing continuous on-site installation tests, Beijing Landsky came up with an ingenious method of installation by using slide tracks. Locating the only installing path within the metal curtain walls that can bypass bolts get in the way. The engineering team cleverly glided the slide tracks mounted with LED lamps into the space between the two layers of the metal curtain walls, making the installation of a large number of LED lamps an enjoyable process.

The LED night lighting scene of the Kunshan Culture and Art Center is a key to the success of the project. Such various lighting "scenes" of elements of "water sleeves", "Southern China ink paintings", "wind", "lotus", "fog" and "water waves" will be staged successively, with the unified theme being an abstracted sense of aesthetics. All related parties such as the builder, owner, architects, lighting designers and LED lamp providers have conducted active communication and collaboration for the lighting scenes carefully, culminating in a universally fulfilling pattern of constantly-changing lighting scenes. With the efforts of all parts, the lighting of this building is granted the second prize of lighting design in 8th China Lighting Award.

—

Material courtesy of: LEOX design partnership, Beijing Landsky Lighting Technology Co., Ltd.

对页：基础照明之余，灯带从暖色到冷色的变幻让纯白色的空间更有趣味。
本页上图：线性灯槽与不同的材料形成有趣的互动。
左图：宁静的发光天棚与戏剧化的场景形成有趣的对比。
下图：灯带沿着顶棚曲线盘旋，与地面的家具相映成趣。

Opposite page: Besides basic lighting, lamps with colors changed from warm to cool make the pure white space more interesting.
Above: Interaction between the linear light troughs and different materials.
Left: Luminous ceiling and dramatic scene together form an interesting contrast.
Below: Lighting grooves and belts on the ceiling interact with the furniture on the floor.

幕墙设计 Curtain Wall Design

在本案中，如何使本质坚硬的玻璃、铝板建材“化直为曲，以刚生柔”，体现了文化艺术中心幕墙设计的特点

AN INGENIOUS USE OF INTRINSICALLY HARD BUILDING MATERIALS AS GLASS AND ALUMINUM PANELS TO ACHIEVE THE EFFECT OF "TURNING THE STRAIGHT TO THE CURVED AND COAXING SOFTNESS FROM HARDNESS" IS A PROMINENT CHARACTERISTIC OF THE CURTAIN WALL DESIGN OF THE KUNSHAN CULTURE AND ART CENTER

设计理念

或许只有在这柔美的水乡，才会成长出如此的建筑，如一枝水中的瑞莲，一缕临风的水袖，曼妙而多姿。建筑师一定是深会了水乡的千古风雅，才在这素墙、曲廊中注入了浓浓的水之神韵，曲之清幽。

初见文化艺术中心的建筑方案，便被其飘逸生姿的外观所吸引。如何将建筑的神韵在表皮中得到最佳的传达，是整个幕墙设计团队所面临的问题。玻璃、铝板本是些生硬的建材，清峻有余而柔美不足。因此，幕墙的构造设计需要做到“化直为曲，以刚生柔”。

建筑与人的交互不仅体现在建筑的整体视觉印象，更多的来自观者内里的体验与感受。一座颇具亲和力的建筑，也定是座舒适的建筑，而建筑幕墙则一定程度上决定了舒适度的等级。因此，如何设计好这层“友好”的界面也是幕墙构造的关键。

设计特点

整个建筑的幕墙系统主要分为三类：组合金属穿孔铝板幕墙、玻璃肋隐索点式玻璃幕墙、陶土板幕墙。

组合金属穿孔铝板幕墙是建筑表达文化元素的重要载体，细密布置的穿孔让素白的铝板幻化成了轻缕纱幔，飘逸水袖。看似随意布置的大小方孔，却在曲面蜿蜒中依稀闪现九仙女们善舞的长袖。此时的铝板已成了灵动画卷。为追求立面的纯净感，铝板幕墙采用了超长板幅的密肋空腔结构，即用飞机铆接工艺将双层铝板固定于内部铝合金骨架上，形成厚度仅80mm的高强中空双层结构，不仅有效隐藏了支撑构件及泛光照明部件，超薄、超长的铝板构造也实现了以刚生柔的设计理念。除了装饰功能，穿孔铝板更是实用的遮阳构件，可有效减少夏季的太阳辐射，降低室内能耗。白天，阳光透过立面穿孔板洒落点点光影，夜晚，剧院内侧的灯影又若隐若现，意趣十足。

文化建筑犹如一个舞台，城市生活便在此处上演，如果说穿孔板犹如帷幕，那玻璃幕墙便是这舞台的中心，视觉的交汇点。因此，通透是玻璃幕墙设计的核心。以最少的结构展示最简洁的空间，玻璃肋隐索点式玻璃幕墙便是最适宜的结构形式。玻璃板缝中内置的不锈钢细索将玻璃面板吊挂于上方结构，百合花造型的不锈钢夹具将面板玻璃与玻璃肋连成整体，抵抗水平风载。超白、高透的LOW-E中空玻璃不仅晶莹通透，更具有优越的热工性能，保证了建筑整体的舒适度。经过精细磨边的超高玻璃肋均采用了彩釉造型，矩形点状图案犹如九仙女们衣袂间洒落的点点灵光，为建筑增添了不少细节之美。

陶土板幕墙主要应用在影视中心四个影视厅的外立面部位。为了突出体量感与节奏感，竖直立面采用平板干挂方式，弧形墙面采用批叠挂接方式，通过板面叠差来贯通直面和曲面间的水平分缝，让立面张力十足、富于变化。正是木色陶土板的点缀，使影视中心成为整个建筑古今传承与变化特色的展示一环，别具一格地呼应存在。

随着传统能源的日益减少，有效利用新能源的重要性愈发突出，节能环保已是成为世界发展主流。在此背景下，光伏技术应运而生，并在建筑工程中得到了较多应用。本工程同时采用了光伏玻璃采光顶及屋顶BAPV，不仅为剧院提供了节能环保的可再生能源，同时也为文化中心树立了“绿色”形象，起到了良好的科技示范效应。

建筑存在的意义不仅满足功能的要求，还是人们升华精神之所。建筑由外而内，通过不同场景的营造，引导人们反复徜徉在具有精神特征的共享空间之中。空间戏剧之美、怀古之悠情，新城之壮丽始终萦绕，挥洒不散。昆山文化艺术中心必将成为昆山最重要的名片，彰显最具特色的城市精神与活力。

—

文：梅春、杨健、计苓 苏州苏明装饰股份有限公司

本页及对页：大剧院及影视中心立面幕墙采用玻璃肋点式隐索玻璃幕墙。玻璃肋隐索点式玻璃幕墙约占工程全部玻璃幕墙面积90%以上，最大跨度16.5m。

This and opposite page: Curtain wall on the facades of the theater and the media center are invisible cable rib-point glass curtain walls that are composed of hollow ultra-white tempered glass, Invisible cable rib-point glass curtain wall which accounts for over 90% of the glass curtain wall area of all the project and its maximum span is 16.5m.

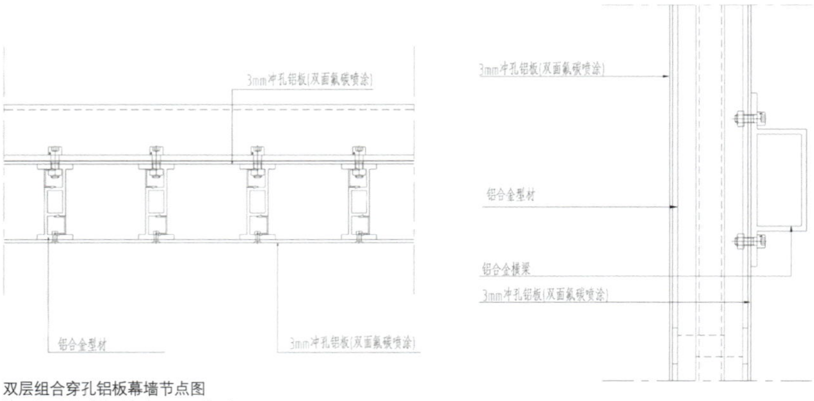

双层组合穿孔铝板幕墙节点图
Panel point of double-layer perforate aluminum plate curtain wall

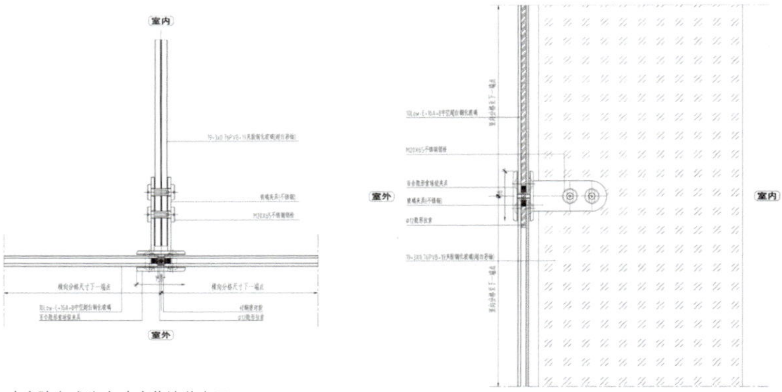

玻璃肋点式隐索玻璃幕墙节点图
Panel point of rib-point glass curtain wall

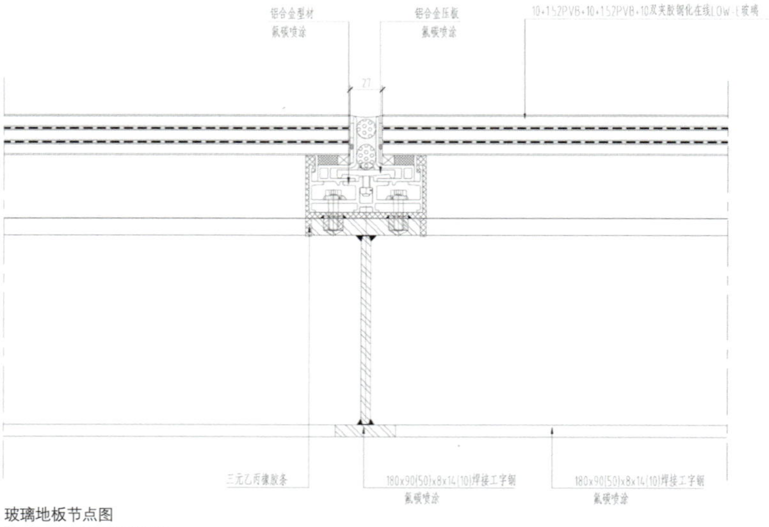

玻璃地板节点图
Panel point of glass floor

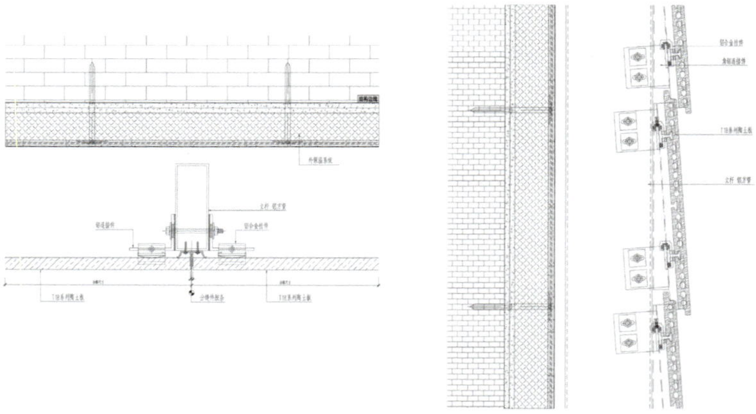

陶土板幕墙节点图
Panel point of clay plate wall

Design concept

Perhaps only in a southern-China watertown can such a strikingly elegant building be born. Like a lotus emerging out of water, or a pair of water sleeves fluttering in the wind, it is such a breathtakingly lithe and graceful presence. Its architect must have fully grasped the quintessential glamour of watertowns, since the structure's unique white walls and winding corridors so brilliantly evoke the allure of the landscape in region south of Yangtza River. Anyone will be captivated by the building's elegant and fluid exterior when taking a first glance at the architectural scheme for the culture and art center. How to best convey the charm of the building through its surface is a challenge faced by the project's curtain wall design team. Since such building materials as glass and aluminum panels are rather hard and inflexible by nature, in the design of curtain wall, the main task lies in "turning the straight to the curved, and coaxing softness from hardness".

The interaction between a building and its visitor does not just mean the overall visual impression the building projects to the visitor, but also the visitor's experience of the building's interior. A friendly building must also be a comfortable building. And to a certain extent a building's curtain wall determines its level of comfort. Therefore, how to achieve an optimal design of this friendly "interface" is essential to curtain wall design.

Design feature

The building adopts three types of curtain wall systems: perforated aluminum composite curtain wall; point-supported glass curtain wall; and terracotta curtain wall. Perforated aluminum composite curtain wall is an important component in showcasing the building's cultural elements. The closely distributed perforations lend an ethereal and delicate look to the whitish aluminum panels, making them look as light as gauze curtains and as fluid as flying sleeves. The seemingly random array of square holes, big and small, magically combine to present a whimsical illusion of flowing sleeves, with aluminum panels virtually turned into a captivating scroll of painting. In order to impart a visual clarity to the facade, the aluminum panel curtain wall adopts a ribbed and hollow structure made of extra-long plates, i.e. using aircraft riveting techniques to fasten double-layered aluminum panels onto the internal aluminum skeleton, creating a high-strength two-layered, hollow structure with a mere thickness of 80mm, which has not only effectively concealed the supporting components and flood lighting components, but also achieved the effect of "coaxing softness from hardness". In addition to decorative functions, perforated aluminum panels also serve as a sunshade that can effectively reduce sun glare and indoor energy consumption. During the day, sunlight will shine through the perforated panels of the facade to create a sparkling effect. And in the night, the lights inside the theater will flicker alluringly, lending a tantalizing, intriguing texture to the exterior of the building.

Cultural buildings are like a stage, on which the show of urban life is presented. If we liken the perforated panels to a curtain, the glass curtain wall shall constitute the very center of the stage, and the visual focal point. Therefore, clarity is key to glass curtain wall design. To showcase space in the neatest way with the barest structure, point-supported glass curtain wall is an ideal system to adopt. Glass panels are hung over the upper structure using stainless steel cables fitted inside the cracks of glass panels, with lily-shaped stainless steel clamps used to bind glass panels and glass ribs together to provide resistance to wind load. Ultra-white, high-permeability LOW-E insulating glass not only offers a striking clarity, but also boasts of an outstanding thermal property, which effectively adds to the building's comfort level. Finely polished extra-high glass ribs are made of ceramic frit glass, with their rectangular patterns dotting the facade in a sparkling manner, lending a charming intricacy to the exterior look of the building.

A terracotta curtain wall system is adopted for the facade of the media center's four auditoriums. To accentuate the sense of volume and rhythm, the vertical facade adopts the dry-hanging technology, with a stacked technology employed for curved walls, using the plate surface dislocation to span the horizontal joint between the straight surface and the curved surface, giving the facade an intense and volatile feel. And it is the wood-colored terracotta plates that make the media center into a unique showcase of a blend of architectural heritage and modern architectural features.

As the conventional sources of energy are dwindling, the importance of making effective use of new energy sources has been steadily mounting. Energy efficiency and environmental protection have become a mainstream theme of world development. It is under such circumstances that the photovoltaic technology emerged, which has since been finding ever wider application in construction projects. This project adopts both photovoltaic glass roofing and BAPV roofing system, which not only provide a renewable energy source, but also help establish a green image for the center.

Architecture is not just meant to serve certain practical functions, it also provides space where people achieve spiritual sublimation. Through its carefully arranged outdoor and indoor settings, architectural structures invite people to stroll through a shared spiritually-enriched space. With the dramatic beauty of space, a haunting nostalgia and the grandeur of the new city seeming to linger perpetually, Kunshan Culture and Art Center is bound to become the most important icon of Kunshan, serving to highlight the charming city's unique spirit and vitality.

—

Text: Mei Chun, Yang Jian, Ji Ling Suming Decoration

对页，从左到右：大剧院外立面的双层组合穿孔铝板幕墙。海角部位玻璃地板采用双夹胶玻璃设计，有效保证玻璃地板的安全性；并且采用在线镀膜处理技术，具有良好的节能性能。

本页：影视中心的陶土板幕墙采用开放式系统；上下板块采用搭接拼缝而成；竖向分缝部位放置通长三元乙丙披水胶条。弧形立面部位陶土板采用披叠挂接方式；保证侧边直面部位陶板与弧面部位陶板水平分格缝标高一致。

Opposite page, from left to right: The double-layer perforated aluminum curtain wall on the outer facades of the theater. Non-slip surface treatment of the glass floor at the promontory (Haijiao) part of the theater is designed to effectively guarantee the safety of the glass floor; and online coating processing technology is employed which enjoys good energy-saving performance.

This page: The thick clay wall of the media center is an open system; the top and bottom plates are consolidated by lap joint; at the vertical parting site, full-length EPDM tapes are placed. Clay plates on the curved facades are joined by shiplap lagging. Clay plates at the lateral facade and the horizontal lattice of the clay plates at the curved sites shall be of the same elevation.

材料研究 Materials Research

穿孔铝板、带肋玻璃幕墙、异型的玻璃纤维加强石膏GRG板材以及无缝艺术环氧磨石地坪构成了文化艺术中心的主要材料特色

THE MAIN MATERIAL CHARACTERISTIC OF THE CULTURE ART CENTER COMPRISES PERFORATED ALUMINUM PLATES, RIBBED GLASS CURTAIN WALL, ABNORMAL-SHAPE GLASS FIBER GRC PLATES AND SEAMLESS ARTISTIC EPOXY TERRAZZO

1

2

1 2 双层组合穿孔铝板幕墙：内外两侧分别为3mm厚的双面氟碳喷涂穿孔铝板、中间设置铝合金支撑型材。

3 穿孔铝板幕墙样板。

1 2 Double-layer perforated aluminum curtain wall: 3-mm-thick double-sided perforated aluminum plates coated with fluorocarbon are installed on both inside and outside, with aluminum alloy supporting structure in between.

3 Curtain wall template of perforated aluminum plate.

3

1

1 3 玻璃肋点式隐索幕墙：由10Low-E+16A+8中空超白钢化玻璃、19+3x0.76PVB+19夹胶钢化超白彩釉玻璃肋、百合隐形索球铰夹具、隐形拉索构成。

2 玻璃肋点式隐索幕墙节点。

1 3 Invisible cable rib-point glass curtain wall: including 10Low-E+16A+8 hollow ultra-white toughened glass, 19+3x0.76PVB+19 laminated tempering ultra-white colored glass ribs, Lily invisible cable ball joint fixtures and invisible cables.

2 Node details of the Invisible cable rib-point glass curtain walls.

2

3

在材料运用中，文化艺术中心结合昆山地域特征和整体设计要求的特点，对材料有针对性地进行选择与研究，并探索出有自身特色的材料加工、研制与施工工艺。

其中在建筑外立面体现水墨肌理的穿孔铝板幕墙、表现出流畅造型的带肋玻璃幕墙和鉴于剧院项目的声学及艺术特点采用的异型的玻璃纤维加强石膏GRG板材以及配合建筑流畅造型的无缝艺术环氧磨石地坪均具有显著特点和独到的施工工艺。

1

2

1 GRG 板细部。
2 顶部为GRG板吊顶，具有非常好的反射性能。
3 观众厅后墙GRG开孔板。
1 Details of the perforated GRG panels
2 High-reflection GRG boards are used to build walls and the ceiling of the auditorium.
3 Perforated GRG panels are also used for the rear wall.

3

穿孔铝板幕墙

穿孔铝板指用纯铝或铝合金材料通过压力加工制成(剪切或锯切)的获得横断面为矩形厚度均匀的矩形材料。考虑到自重、耐久性、表面无焊点，此次选用了西飞集团公司作为配合单位，用制造大飞机机翼的技术，实现了设计要求。最大的一块铝板长度达到9.6m，铝板共1000余块。该铝板幕墙不仅可以通过折弯、冲压成型，表现出艺术中心的飘逸流畅造型，还通过穿孔与颜色变化，完美地体现了独特的水墨肌理。部分幕墙铝板长达9m，在运输、安装过程中必须保证铝板的完整性，所以会增设专用运输车，吊装过程也需要在风速在五级以下分部逐次进行。

隐锁带肋全玻幕墙

项目中的玻璃肋点式隐索玻璃幕墙由中空超白钢化玻璃、夹胶钢化超白彩釉玻璃肋、百合隐形索球铰夹具、隐形拉索构成。在玻璃面板板缝中设置有隐索；通过索与夹具的配合，避免玻璃重力及其产生的偏心力传替到支撑玻璃肋结构上，减轻了玻璃肋孔部位集中力负担，提高了幕墙系统的安全可靠性。

GRG材料

鉴于剧院项目的声学及艺术特点，本工程顶棚和墙面大量采用了异型的新型装饰材料玻璃纤维加强石膏GRG板材。GRG即玻璃纤维加强石膏板，它是一种特殊改良纤维石膏装饰材料，造型的随意性使其成为要求个性化的建筑师的首选，它独特的材料构成方式足以抵御外部环境造成的破损、变形和开裂。GRG材料良好的声学性能，适用于剧院。在艺术中心的观众厅顶棚顶面为花瓣形的GRG拼接而成，墙面为不规则带方孔扩散体GRG

1

2

3

1 舞台墙面两种穿孔率吸声结构，面板为8mm厚穿孔硅钙板，内衬无纺布及环保型吸声材料。

2 木开槽板吸声结构，木板外侧面开槽，内侧面开孔，形成穿孔板，板后衬无纺布及环保型吸声结构，该吸声结构用在声闸、后台出入口等处。

3 电影厅木丝板吸声结构：木丝板具有良好的透声性能，与内部环保型吸声材料共同构成性能良好的强吸声全频吸声结构。

1 There are two types of sound absorption structure for the walls on stage, with different perforated percentages; the panel is 8 mm thick perforated calcium silicon plate, lined with environmentally non-woven fabric.

2 Sound absorption structure of wood notched panel, planks are slotted outside the side face, punched inside, formed as perforated plates, with the lining of non-woven fabric and environmental sound absorption materials, which are placed at backstage entrance etc.

3 Sound absorption structure of excelsior plates in film hall: excelsior plate has good acoustic performance. Together With the internal environmental absorption materials, they form strong absorption structure of all frequency sound.

拼接而成。

大面积、高空间、弧形、异形的GRG的运用，使得本项目施工难度尤其突出。以现场放线为例，专业厂家根据图纸的曲线要求，用专业的犀牛软件进行建模、分块，把整体的一个面分成若干个小块。以小块为一个单位，进行GRG的分块施工。现场则根据图纸曲线的走向和尺寸进行1:1定位放线、钢架的制作，根据专业的施工方法，再把分块的GRG依照犀牛软件的图示要求进行拼接。

环氧磨石地坪

为完美诠释文化艺术中心流动唯美的整体效果，地面铺装材料选用了新型材料——环氧磨石，较之普通天然石材，其具有无色差、整体无缝、图案流畅、附着力强、耐冲击的效果。其高雅华贵、晶莹剔透，表面光洁度能达到的镜面效果，超过中高档大理石的装饰效果。

用于地面铺装材料的无缝艺术环氧磨石地坪主要成膜物质采用低黏度环氧树脂和自制的环氧加成改性的环脂胺类无溶剂固化剂加磨石骨料（特质彩色陶瓷颗粒、高透玻璃镜子、天然彩砂）制成；高装饰性环氧磨石地坪的磨石集料是实现装饰性效果的主要载体，坚硬耐磨、色彩鲜艳、搭配得当，使创意呈现淋漓尽致。该地坪附着力极好，整体无缝易清洁、耐冲击，经久耐用。更因其含镶嵌式图案，是该地坪的艺术观赏性和实用性得到了完美结合，图案稳定不易磨蚀。该地坪引进美国宣伟公司的先进技术，并经过研发人员工艺改进和技术提高，利用先进的“玻化”技术从根本上解决了水磨石地面的缺陷。在施工环氧磨石浆料时，利用激光找平仪对地坪进行激光找平，可以使浇注的环氧磨石地坪达到很高的平整度，整体误差不超过1厘米。

此外，作为剧院，必不可少的声学设计，舞台设计等，其重要环节中的舞台、观众厅以及电影厅墙面的吸声结构材料也各具自身特色。

1

1 环氧磨石地坪样板。
2 3 环氧磨石地坪实景效果。
1 Epoxy terrazzo floor templet.
2 3 Views of Epoxy terrazzo floor in theater.

2

3

To material application, combined with the local features of Kunshan and characteristics of the whole design requirements, the culture art center particularly selects and researches the material, and has found the unique technologies for material processing, development and construction.
Wherein, obvious characteristics and unique construction technologies are gifted to the punched aluminum plate curtain walls which display ink painting textures on the outer vertical surface of the building, the ribbed glass curtain walls which show smooth models, the abnormal-shape glass fiber reinforced plaster GRG plates with acoustic and artistic characteristics for the theater project and the seamless artistic epoxy terrazzo terraces fitting the smooth models of the building.

Perforated aluminum plate curtain wall
The perforated aluminum plate is the rectangular material of which the cross section is rectangular and has uniform thickness, and is made by pressurizing (cutting or sawing) pure aluminum or aluminum alloy material. Considering the self weight, durability and no welding point on the surface, it selects XAC group as the cooperative institute, and realizes the design requirement by the technology for manufacturing wings of big planes. The biggest aluminum plate is 9.6 meters long, and there are more than 1000 aluminum plates. The aluminum plate curtain wall can be formed by being bent or punched. It not only displays the elegant and smooth model of the art center but also perfectly displays unique ink painting texture by the variation of punctures and colors. It can perfectly display unique ink painting textures. Some aluminum plates of the curtain wall are 9 meters long. In transportation and installation, it must guarantee the integrity of the aluminum plate, thus, special transportation vehicles are utilized, and the hoisting needs to be carried out by parts and times when the wind velocity is lower than level 5.

Hidden-lock ribbed full-glass curtain wall
The glass rib-point type hidden-lock glass curtain wall of the project comprises hollow super-white tempered glass, doubling tempered super-white colored glaze glass ribs, lily hidden cable-ball hinge clamps and hidden backstay cables. Hidden cables are set in seams of glass surface plates. Because cables and clamps are fitted, glass gravity and eccentric force generated by glass gravity are prevented from being transmitted to the support glass rib structure, the burden of concentrated force at the glass rib holes is relieved, and the safety and reliability of the curtain wall system are improved.

GRG material
Because of the acoustic and artistic characteristics of the theater project, a great quantity of abnormal-shape novel decorative material (glass fiber reinforced plaster) GRG plates are applied to the ceiling and ground of the engineering. GRG means glass fiber reinforced gypsum plates, which are the specially-improved fiber gypsum decorative material. The random model makes it be the first choice for the architect who requires individuality. Its unique material composing way can sufficiently resist the damage, deformation and crack caused by the outer environment. The excellent acoustic property of GRG material is suitable for theaters. The ceiling top surface of the auditorium of the art center is formed by splicing GRG, and the wall surface is formed by splicing irregular diffusors GRG with square holes.
The construction of the project obviously becomes difficult because of the application of large-area, high-space, arc-shaped and abnormal-shape GRG. Take the on-site paying off as an example, as required by curve requirements of the drawing, the professional manufacturer models and blocks with professional Rhino Software, and divides one whole surface into a plurality of blocks. One block is one unit, and then GRG blocked construction is carried out. At the site, 1:1 positioned paying off and the manufacturing of the steel frame are carried out according to the trend and size of the drawing curve. According to the professional construction method, GRG blocks are spliced as the requirements of the diagram of the Rhino Software.

Epoxy terrazzo
To perfectly explain the flowing and graceful comprehensive effect of the culture and art center, the novel material which is epoxy terrazzo is used for paving the ground, and compared with the common natural stone, it has the effects of no color difference, no integral seam, smooth patterns, strong adhesion strength and impact resistance. It is elegant, luxurious and transparent. Its surface glossiness can get the mirror effect which is better than the decorative effect of the middle and high-grade marble.
The main film material of the seamless artistic epoxy terrazzo used as the material for ground pavement is mainly made by mixing low-viscosity epoxy resin, self-made epoxy improved alicyclic amine solvent-free curing agent, terrazzo aggregates (special colored ceramic granules, high-transparent glass mirrors, natural color sand). The terrazzo aggregates of the high-decorative epoxy terrazzo terrace are the main carriers for realizing the decorative effect. They are hard, abrasion resistant, bright and properly fitted so as to perfectly display the originality. The floor has outstanding adhesive force, no seam, easiness for cleaning, impact resistance and durability.
In addition, it contains embedded patterns so that the artistic ornamental performance and practicality of the terrace are perfectly combined, and the patterns are stable and corrosion resistant. The advanced technology of America Sherwin-Williams company is introduced for the terrace, and technically improved by R&D personnel, and advanced vitrification technology is used for essentially solving the defects of the terrazzo ground. When sizing the epoxy terrazzo in construction, the laser leveling instrument is used for the laser leveling of the terrazzo, and the pouring epoxy terrazzo terrace can get high flatness, and the whole error is not higher than 1cm.
In addition, as the necessary acoustic design, stage design, etc., the acoustic absorbing structural materials of the stage, and the wall surfaces of the auditorium and movie hall are variously characterized.

2009年10月19日，昆山文化艺术中心开始桩基施工。
On October 19, 2009, Kunshan Culture and Art Center began the construction of pile foundation.

2010年7月10日，昆山文化艺术中心进入土方开发阶段。
On July 10, 2010, Kunshan Culture and Art Center was in the construction phase of pit supporting.

基坑施工 Foundation pit construction

基坑开挖面积大、维护结构长、开挖深度深且场地土质含水率高，因此难度较大。在施工方案中对于其挡土系统、止水系统、支撑系统、栈桥系统等多方面进行了专门考量，并实时监测。

Under the condition of high moisture content of soil ground, it was difficult to pit such a large and deep excavation area and long maintenance structure. For the construction scheme, special considerations were provided in terms of the soil retaining system, water system, support system and pier system, real-time monitoring

2011年12月22日，影视中心二次结构完成。
On December 22, 2011, secondary structure of Media Center was finished.

2012年9月7日，大剧院海角钢结构施工现场。
Construction site of the steel structure of auditorium on September 7, 2012

2011年12月22日，大剧院结构完成。
On December 22, 2011, the structure of auditorium was accomplished.

结构相关难点 Difficulties and challenges

结构施工期间，专门针对支模架的技术难点工程进行了攻关。为解决振动、撞击等产生的声干扰问题，除了对上部设备及管道所作的减振隔振措施，采用了浮筑楼面减振隔声的结构设计方式。

During the construction phase, there were special solutions for technical difficulties of formwork supporting. To solve the sound interference caused by vibration and collision, besides isolation measures on the equipments and pipelines, floating floors were adapted to [illegible] vibration isolation.

2012年7月6日，屋面飘带幕墙施工。
On July 6, 2012, the curtain walls with ribbon shape on roof were in progress.

2012年6月15日，影视中心外陶土板保护层施工。
On June 15, 2012, protective coating of clay plate on Media Center was under construction.

2011年1月7日至1月11日，昆山城投公司与中建院及上海和甲幕墙设计咨询有限公司相关人员前往武汉凌云军品八车间，对本项目金属幕墙观察样板的加工制作的全过程进行了现场监察与跟踪。
On January 7 to January 11, 2011, Kunshan City Construction Investment Development Company and Shanghai DHD Curtain wall Design & Consulting Company sent related personnel to the plant and monitored the whole process of making the observation samples of metal curtain walls.

幕墙试验 Curtain walls trial

铝板幕墙采用了超长板幅的密肋空腔结构，即用飞机铆接工艺将双层铝板固定于内部铝合金骨架上，实现了支撑尽可能弱化，并充分表现飘逸感及昆山文化内涵的目标。

Aluminium curtain walls have super long hollow-ribbed structure, that is to say, double-layer aluminum plates are fixed on the internal skeleton of aluminium alloy by aeroplanes. This approach softened the visual effects of supporting structure as much as possible and elegantly expressed the profound Kunshan culture and history.

2012年3月10日，大剧院舞台设备安装。
On March 10, 2012, the equipment installation of auditorium stage started.

2012年3月10日，舞台设备安装现场，施工人员正在焊接作业。
On March 10, 2012, workers were welding on site.

舞台设计与安装 Stage design and installation

项目的核心功能区是舞台，舞台基坑加地上高度共达50m，是个类似筒的空间。在设计与安装的配合下，布置整齐划一，便于技术维修人员工作及通行，可满足不同演出需求。

The core area of project is stage, the height of both ground and underground of foundation pit is up to 50 meters together, similar with a tube space. In conjunction with the design and installation, the layout is in order to facilitate technical maintenance and passage of personnel, which satisfy different performance requirements.

崔愷，中国建筑设计研究院（集团）建筑设计总院
名誉院长、集团副院长、总建筑师

Cui Kai, honorary president, vice president and
chief architect of Architecture Design
General Institute of China Architecture Design &
Research Group

Interview with Cui Kai

Rational Thinking About the Perceptual

关于感性的理性思考·崔愷

新印象：文化艺术中心设计思路是怎么形成的，构思过程有没有什么故事？或特别值得回顾的环节？

崔愷：当时我正在朱家角设计一个酒店项目，《时代建筑》支文军主编邀请我对昆山文化中心项目的方案提提意见。我那是第一次到昆山，对城市还不是太了解，通过市规划局的介绍，了解了这个项目的背景和相关情况后，提出了一些初步设想，我认为这个项目不应该是一个孤立的建筑，它地处体育中心旁，应该与体育中心和周边建筑形成有机的整体；其次它邻近森林公园，是城市绿化系统的一部分，应该是公园中的建筑；此外场地中有一条河，它应该有更多的亲水空间，表现江南水乡的传统特征。我的这种观点得到了业主和相关专家的认可，后来昆山市规划局徐振敏局长还亲自来北京邀请我重新设计这个项目，当时我觉得让原来的设计单位修改更合理，但他们执意坚持，我也只好“从命”了。

由于有了前面的过程，我们对自己也提出了更高的要求，不只是做个漂亮的项目，而希望更多地解读这个城市的文化。应该说我们以往大多数的设计多从理性出发，由内向外，从功能到造型等来考虑设计。而这次相反，在特定环境中设计这样一个项目，我们有一种感情在里面，有一种讴歌赞美这座城市的情怀。从这座城市的长远规划、生态环境、历史文脉和水乡特色中，从外在的线索寻找设计的思路，通过感性的方法，外部分析的方法产生的这个项目构思可以说是由外向内的，也反映出了这个建筑的特殊性。可它虽然源自浪漫的构思，我们却不想仅仅堆砌一些形体或漂亮的效果图来取悦业主，而更要考虑如何把构思贯穿到整个空间，和功能的设计要很自然地结合。为此我们花了几个月的时间从外部的形态到内部的组合，不断修改和深化方案。我们希望营造的是一个开放性的建筑，与周边环境相互关联。这种大体量的建筑最大的问题就是很容易沉闷，孤立，甚至高高在上，比如剧院的庞大的舞台部分是内部使用的，容易造成封闭感。我们在前面设置开放性入口，透明的观众厅将人流引来，让后台区域和另一个广场结合起来，并把影视和会议中心的入口设在这里，穿过它，是滨水的市民广场。这样的设计使这个建筑没有消极的背面。并且通过连廊与体育场、体育馆有机的连接在一起，体现了城市建筑空间应有的共享性和协调性等，也体现了对周围建筑和其设计者的尊重。

还应特别提到的是我们遇到了很好的业主，城投公司的周总对我们设计团队的全力支持，现场指挥的范总与我们在每项设计中的沟通都及时而充分，我们也因此能更加深入地控制每一个环节。另外我们还承担了室内设计工作，与法国的艺术设计家阿兰博尼(Alain Bony)积极合作，不断探索，完善我们的构想，整个过程是一个辛苦又愉快的过程。

新印象：这个项目都有哪些超出常规的设想？通过什么方式解决的？

崔愷：占建筑立面面积五分之三以上的金属幕墙是项目的一个重点与难点，不仅要表现设计意向上大的视觉效果，同时还希望有细节、有变化、有质量要求。参考国内外的不同案例，我们决定采用像素化穿孔立面图形的做法，利用穿孔幕墙的阵列疏密变化将昆曲表演艺术照片抽象地刻印在立面上，如山水长卷，浓淡相宜，将来配上灯光，可以呈现动态的效果。另外为了透光和与室内空间互动，

幕墙是双层双面的，因此对它的吊挂工艺都做了很细致地研究，多次做样板，从而保证了室内外观看的效果。在照明上，尽管穿孔设计为LED照明系统创造了安装条件，出于节能的考虑，我们建议不要过多过亮。结合外部投光灯等，让建筑夜间照明呈现出像纱幔一样的柔和效果。这部分工作目前正在调试，最终完成效果还要拭目以待。

新印象：这个项目实施希望给这个地域带来哪些变化？在这里力图营造的气场是怎样的？实现效果如何？

崔愷：我们希望在这营造的是一种轻松氛围，建筑是淡雅的、流畅的造型；景观应与环境融为一体，提供更多的室外共享休闲空间。我们希望利用原有的河道在建筑旁营造亲水空间，有长长的敞廊在茂密树木中游走，背后隐约可见的是剧院建筑，在这里人们可以谈恋爱、休憩、娱乐，水中映出建筑的倒影，甚至可以看到栖息在这里的鸭子水鸟，一片新水乡的田园风光。我们为此作了相应的景观概念设计。但是后来的景观设计由其他人做，对这些设计思路有所改变，代之以不太生态的浅水池及大面积铺地广场，做得很生硬。我们提出过意见，但那位设计师比较固执，说河水水质不太好，所以不能直接用作景观水，但我想通过改良水体创造亲水空间，不仅可以解决景观问题，对整个城市环境都将有更积极意义。我们希望后期要整改，增添更多的植被，改良水体。毕竟“水”也是我们建筑创作的主题呵！

新印象：通过文化艺术中心的设计，这样的建筑群想为昆山市制造哪些可能性？

崔愷：昆山这个快速发展的工业城市不再满足于GDP的逐年上升，建造优美的、有文化气质的城市应该说是一个更保本，更长远的追求。我们也在考虑这个城市的气质，水乡的白墙黛瓦，昆曲的意蕴，甚至其特有的吴侬软语都是重要的传统文化元素，所以把柔、轻、淡雅作为它的定位。我们不敢说文化艺术中心的建成对这座城市会有怎样的影响力，但我们想让它成为大家欣赏的建筑。影视中心、多功能剧场、大剧院，不同的功能可为人们提供多种文化消费的场所。它不应是一个高高在上的艺术殿堂，而是充满活力的，有人情味的城市居民文化生活的有机组成部分。事实上在我们设计、施工期间，周围的住宅楼群正逐步建成，对面的办公大楼也正式投入使用，随着城市的发展，这一区域将会成为城市副中心，相信文化艺术中心作为这一整体规划的有机组成也会越来越发挥更充分的服务功能及推动作用。听说这个刚刚落成的建筑近期被市民选入苏州市十大优秀公共建筑，让我很意外很欣慰。

新印象：您认为在目前，国内的城市、社区与属于当地的文化建筑和设施相互关系应该是怎样的？

崔愷：一个文化建筑应该融入这个城市的生活。很多城市的政府在做文化建筑的时候往往把它放在一个新区，孤立在大广场中央，虽然成为了所谓标志性建筑，但明显脱离了城市生活，市民难以接近，没有大型演出时也没有其他经营的空间，曲高和寡，日渐冷清成了这类建筑的特性。

所以，我每次设计完成后，并不满足于一些静止的漂亮照片，而是希望过若干年后回到那里，看到它跟城市生活密切地联系在一起了，这才是我最希望看到的。

New Impression: How was the design mentality for the Culture and Art Center developed? Any stories to share in the course of conception?

Cui kai: At that time I was designing a hotel project when Zhi Wenjun, chief editor of *Time Architecture* invited me to give suggestions on the scheme of Kunshan Culture and Art Center. It was my first time in Kunshan. I still knew little about the city. Through the introduction by the planning bureau of the city, learning about the background and other related conditions of the project, I put forward some preliminary ideas. I think that the project should not be an isolated structure. It' s located near the Sports Center, so it should form an organic whole area with the Sports Center and surrounding constructions; besides, it' s close to the Forest Park, as a part of the urban greening system, it should be a structure in the park; there' s a river in the area, it should be of more hydrophilic space, so as to express the traditional characteristic of a watertown. My ideas were accepted by the owner and related experts. Afterward, Xu Zhenming, director general of Kunshan Urban Planning Bureau came to Beijing to invite me to re-design the project. At that time, I deemed it more reasonable for the original designing unit to modify the design. However he persisted his request, I had no other choice but to follow the request.

With the above-mentioned process, we raised a higher requirement for ourselves. Our aim was not simply a nice-looking project, but to better interpret the culture of the city. Most of our previous designs started from rationality and considered the design from inside to outside and from function to style. To the contrary, this time the project will be designed under a special environment. We have some affection in it and wish to eulogize the city. From the long-term planning, ecological environment, history and watertown characteristic of the city, from the external clues, we're searching for a design mentality. The project design idea developed by means of perception and external analysis may be outside-in, which reflects the particularity of the architecture. Though it stems from a romantic conception, we do not aim to please the owner by piling up some forms or presenting a beautiful design sketch. Instead, we'll consider how to run our conception throughout the space and achieve a natural integration with the functional design. For the purpose, we have spent months of time in modifying and deepening our scheme from external form to internal combination.

What we wish to create is an open architecture, inter-related with the surrounding environment. The biggest problem for such a large architecture is that it's easy to look oppressive, isolated or even aloft. For example, the huge stage of the theater is for internal use, which is easy to produce a feel of being closed. We set an open entrance at the front, so the transparent audience hall guides in the flow of people, the backstage area is connected with another square, and set the entrance to the video and conference center at this position, go through it, you' ll arrive at the waterside citizen square. Such a design rids the architecture of any negative back face. It' s connected with the stadium and gym via a corridor, which reflects the sharing and coordination characteristic of an urban architectural space, and respect for surrounding constructions and their designers.

What should be especially mentioned is we have met a good client. Zhou Jichun, chairman of Kunshan City Construction Investment & Development Company was fully supporting our design team. Manager Fan who commanded on the site made a timely and full communication with us on each design item. So we were able to control each link more deeply. Besides, we also undertook the indoor design work, actively worked with Alain Bony, artistic designer from France, kept exploring to perfect our conception. The whole process was hard but pleasing.

New Impression: Are there any unusual conceptions in the project? How are they tackled?

Cui kai: The metal curtain wall which takes up over 3/5 the area of the building facade are a key and difficult part in the project, which not only should express a huge visual effect in design intention, but also aim to reflect minutia, change and quality requirement. By referring to different cases at home and abroad, we decided to adopt the practice of pixelated perforated facade pattern, in which the array density change of the perforated curtain wall is employed to imprint the photo of Kunqu opera performance art onto the facade, like a landscape roll which is appropriate in shade. In the future, with lights, it can display a dynamic effect. Besides, for the purposes of light admitting and interaction with the interior space, the curtain wall is double layer and double side. So its hanging workmanship is studied in detail, with samples prepared for times, so as to ensure the viewing effect from indoors and outdoors. As to the lighting, though the perforation design has created installation condition for LED lighting system, for consideration of energy saving, we suggest it should not be too many and too bright. Together with the external projectors, etc., it can make the night-time lighting of the architecture present an effect as soft as gauze curtain. The part of work is currently under debugging, so let's wait and see the final effect in completion.

New Impression: What did the project aim to bring to the area? How about the atmosphere created here? How is the effect achieved?

Cui kai: We wish to create a casual atmosphere here. The architecture is of a simple and smooth model; the landscape should integrate with the environment to provide more outdoor sharing space for relaxation. We wish to create a hydrophilic space at the architecture by making use of the original watercourse, with a long corridor stretching in the dense trees, behind which is the theater visible indistinctly, where people can date, have a rest and amuse themselves, where the water reflects the reflection of the architecture, and you can even see ducks and water birds, what a idyllic scenery of new watertown! So we have conducted a landscape concept design. However, afterwards the landscape design was undertaken by others who made certain changes to the design mentality, in which a shallow pool not so ecological and a spacious paved square were adopted instead, very unnatural. We have aired suggestions, but the designer was a bit stubborn, insisting that the water quality of the river was not so good, so it cannot be directly used as landscape water. But in my view, the improvement of water to create a hydrophilic space not only can solve the landscape issue, but also is of more positive significance to the whole urban environment. We hope that modification should be made in later phase, to increase more vegetation to improve the water quality. After all, "water" is also a theme of our architectural creation.

New Impression: Through the design of Culture and Art Center, what possibilities will the architectural group create for Kunshan City?

Cui kai: Kunshan, such a rapidly developing industrial city, is no longer satisfied with the year-on-year rise of GDP. Building a beautiful city with cultural disposition should be a more value-preserving and long-range pursuit. We're also considering the disposition of the city. The white walls and black tiles of the watertown, the implication of Kunqu operas, or even the special dialects are all important traditional cultural elements of the city, so we position it as soft, light and elegant. We dare not say what influence the completion of the Kunshan Culture and Art Center will bring to the city, but we want it to be an architecture appreciated by people. The media center, the multi-function theater and the grand theater provide a place for the various cultural consumption of people. It should not be an aloft art palace, instead it should be an organic part of the urban citizens' cultural life which is filled with vitality and human touch. In fact, during the period of our design and construction, the residential building groups around the area are being completed while the office building on the opposite side is also officially put into use. With the development of the city, the area will become a sub-center of the city. We' re convinced that Kunshan Culture and Art Center, as an organic part of the overall planning, will increasingly bring into play its service function and driving force. I feel very surprised and happy that the architecture completed recently was selected by the citizens among the top ten public architectures in Suzhou.

New Impression: In your view, what kind of a relation should be between cities and communities and the local cultural architectures and facilities?

Cui kai: A good cultural architecture should blend into the life of the city. The governments of many cities usually place a cultural architecture in a new district, isolate it at the center of the huge square. Though it has become the so-called landmark architecture, it's obviously out of the urban life, hard for citizens to get close. It has no other operation space if there's no large performance. It's gradually disserted, which is the common fate for architectures of this kind.

Therefore, each time after I finish my design, I am not satisfied with the motionless beautiful pictures, instead I hope that after years of time when I revisit the architecture, I can find it closely integrated with the urban life. That's what I expect.

阿兰·博尼，法国建筑造型艺术家，昆山文化艺术中心首席色彩顾问

Alain Bony, French Architectural Artist, chief color consultant to Kunshan Culture and Art Center

Essay of Alain Bony

On the Palette and Artistic Aura

色彩及艺术氛围阐释·阿兰·博尼

文化与经验的融合

习惯于三十年来与努维尔跨越时间空间的合作，而随安德鲁和国家大剧院进入中国，也已十年之久。于我而言，与建筑大师崔愷先生在昆山的合作，是一个十分愉悦的过程，甚或，一种精神的享受。

与崔愷先生的合作，我发现了另一种对建筑高雅与飘逸的体悟：行云流水、举重若轻，十分值得体味。让我也有很多新的触动，愿意将我的文化与经验与之深厚的积淀相融合。

昆山项目的甲方，给我们留下深刻的印象。整个团队，尤其是领导人的专业性与敬业精神，让我们由衷为昆山感到幸运，也看到中国建筑时代的到来。毕竟，没有任何一个顶级的建筑作品，可以离开一个严格而智慧的业主。

逐层深入的介入

这个项目中，我们的介入逐层深入：从公共空间地面和墙面的色彩确定，到大剧院和多功能厅的色彩选择及艺术氛围渲染，直至室内艺术投影作品的实现。秉承一贯的理念：让建筑本身成为艺术品。而一切艺术手段都基于对建筑本身的理解："建筑即是一段旅程。"

人们对于建筑的旅程，由远及近、由外及内，渐入佳境……曲径通幽，引领我们穿过中国南方的竹径与水景，一片宁静祥和的氛围。

外立面镂空层叠，淡雅含蓄，造型舒展，富有变化。不同的建筑体块，大剧院与影视中心，自然流畅地编织在一起，信步徜徉，移步换景，构成闲逸而有情趣的外部空间。

大剧院前厅雕塑手法初现。墙体造型自由、优雅、流畅、富有韵律，秉承昆曲"水袖"清丽淡雅之风，行云流水，一气呵成。在色调上，色差甚微的暖色木沿墙体流线交替分布，融入整个空间平和淡雅的氛围。而淡灰与乳白地面，在夜晚灯光下，熠熠闪现。

投影艺术作品取自昆曲的元素与色彩，遵循建筑师对唯美意境的追溯，透过艺术家的视角与演绎，浮动于白色墙面，犹如瓷器般的效果。隐逸于建筑空间之中，灵动飘忽，若隐若现。观者静心漫步，不断发现、赏析，更添雅兴……

每种色彩寓意不同

红色：热烈隆重，华丽而激动人心。紫色：高贵典雅而略带神秘。同时也是当地历史文化精粹之昆曲色彩，在现代灯光中的演绎。暖金：雍容而温暖，深邃而耐人寻味。金色：平静而中性，寓意辉煌与丰盈。

玫瑰花顶：雕塑手法再现，愈加具象与细腻。经过甄选的红色、紫色、金色与暖金色，使大剧院观众厅的"花顶"呈现不同的艺术特质，为各样演出与场合，营造与之切合的情感氛围……而场前、场间与场后场景的设置，色彩与亮度渐变，为观者心境的准备，做好铺垫。

淡色的侧墙，宛如平静展开的画布，任由花瓣溢出的色彩，洒落其上，充盈整个空间。浮动其上的投影，缓缓流动，渲染"轻雾笼罩"的朦胧氛围。

如此手法，由天顶的光影与色彩，依不同演出、事件与用途，渲染表情丰富的艺术氛围，远非单纯墙面涂层可以实现，在大剧院作品中罕有挖掘。

从某种意义上讲，观众厅空间本身即是一种场景设置亦或一件光影艺术作品。伴随色调、照度和节奏的变换，空间被赋予不同的情感。引领观者去发现、惊喜与期待。建筑之旅也随之进入高潮。

A fusion of culture and experience

Accustomed to conducting time-and-space-transcending collaborations with Jean Nouvel for the past 30 years, I' ve been in China for nearly a decade since first arriving here with Paul Andreu for the National Centre for the Performing Arts project. For me, working with Cui Kai, a prominent architect in China, was a highly enjoyable experience, indeed a spiritually fulfilling one.

In collaborating with Cui, I came to a renewed appreciation of architectural elegance: architecture may demonstrate such a smooth transition and fluidity and such a magical blend of concreteness and ethereal grace that I couldn't help but feel greatly inspired. I'm now so willing and eager to integrate my cultural understanding and experience into this unique, richly nuanced architectural style. The client of the Kunshan project also greatly impressed me. The professionalism and dedication to work of the whole team, especially the leaders, convinced me of the picture of the golden era of Chinese architecture. I am so glad for Kunshan. After all, no architectural work, however brilliant, can survive without an exacting and inspired client.

A progressive designing approach

In this project, the design approach we adopted was a progressive one: throughout the design process, from the determination of hues for the floor and wall of public space, the selection of colors and creation of artistic atmosphere for the grand theater and multi-functional hall, to the realization of indoor projection art works, we strove to pursue one unified goal-- to turn architecture itself into a piece of art. All artistic means are based on our understanding of architecture: architecture is a journey.

In appreciating a piece of architecture, people always approach from afar, and from outdoor to indoor, with the brilliant qualities of the delicate structure gradually blossoming along the way...

A meandering path leads us through bamboo groves and water scenes so typical of Southern China, exuding a perfect aura of peace and serenity.

The layered, "openwork" facade presents a neatly elegant and nuanced look, featuring a fluid and dynamic form rich in subtle changes. Varied architectural components, including the grand theater and the media center, are organically woven together: take a leisurely stroll through it, and you will be presented with changing views as you move along, with a relaxing and intriguing external space brilliantly created.

In the lobby of the grand theater, some intricately-fashioned sculptural elements are sure to delight you. Elegant, fluid, and rhythmic, the body of the wall conveys a soothing sense of freedom, strongly evocative of the Water Sleeves, a feature of the Kunqu Opera costumes that typifies elegance and fluidity. The warm-hued wooden elements with the slightest color differences are distributed along the body of the wall in a fluid and alternate fashion, effortlessly melting into the aura of quiet elegance pervasive in the whole space. The light brown and white floor also gives off a subtle, glittery sheen in the night light.

The project art works used in the project also draw upon the elements and hues of the Kunqu opera. In our pursuit of a purely aesthetic mood and by employing the inspired perspective and techniques of artists, we manage to create a wonderful porcelain-like visual effect on the surface of the wall, a visual feast that shows its tantalizing presence throughout the space, with its alluring ethereality adding considerably to the glamor of the interior. Audiences are advised to slow down and appreciate the mesmerizing, nuanced light effects, which will surely make their visits so much more fulfilling.

Implicit meaning of colors

"Rose-petal" ceiling: a sculptural treatment is given to the ceiling too, lending it a rich visual intricacy. A carefully selected palette of red, purple, golden and warm golden hues also helps affect varying artistic auras on the ceiling to suit the different emotional settings necessitated by different shows and performances, with the stage setting and gradual changes in color and brightness also serving to prepare the audiences for mood change.

The light-colored side wall looks like a quietly unfolding canvas, onto which the hues of the petals wantonly spill, filling the entire space. The projection effects fluttering above flow slowly, creating a unique mist shrouded hazy atmosphere.

The technique of allowing the lights, shadows and hues of the ceiling to affect richly varied artistic auras to suit different occasions, settings or purposes is far superior to the mere adoption of wall coatings, and is also rarely pioneered before in the realm of theater.

In a sense, the space of the auditorium itself is a part of the stage sets or a piece of light art, with the changing hues, brightness and rhythm instilling different moods in the space, leading the audiences to discover, to expect and to get delighted, thereby bringing their "architectural journey" to a climax.

李跃文，昆克剧院工程咨询（北京）有限公司副总经理

Li Yuewen, vice general manager of Kunkel Theatre Engineering & Consulting (Beijing) Co., Ltd.

Interview with Li Yuewen

Rigorous and Challenging Design

严谨与充满挑战的设计·李跃文

新印象：您个人觉得文化艺术中心给您以什么样的感觉？它的特点如何？

李跃文：其实每个剧院的功能都大同小异，对于我们来讲，在设计之初首要的工作是与使用方（或投资方，以下简称业主）做好深入的沟通，充分了解业主的愿望，也就是对待建剧院的使用要求，即功能定位，对于设计者来讲，了解剧场的功能定位是十分必要的，因为这是设计剧场的根据和前提。对于昆山文化艺术中心来讲，这是一个高雅又不失亲民的艺术殿堂，其地理位置优越、交通便利，是昆山市城市格局的重要组成部分。整个建筑由国内著名建筑大师担纲设计，结合了昆山地方文化特色，是一个地方传统文化与建筑艺术的完美结合体，是一个处处流露着地方传统文化的建筑艺术精品。

新印象：该舞台设计有哪些难点和特殊要求？在国内外是否有先例？戏剧舞台的设计要求与西方剧院的设计要求有哪些不同？

李跃文：对于舞台设计来讲，其难点可能与建筑设计类似，也是要将现代的舞台技术与传统的艺术文化结合，也就是既要能够实现中国传统剧目的演出要求，又要能够满足西方传统剧目、现代流行剧种以及未来的各种可能，我们所要做的就是了解中国戏剧对舞台的要求，同时结合现代舞台技术，对于我们和业主方来讲这是一个多次的沟通过程，也是一个学习过程。

近年来，剧院舞台设计产生了巨大变化，更宏大、更精细、超乎寻常的表演要求比以前更多、更优质的舞台设备，需要将更新、更高、更成熟的技术运用到舞台设施和设备的设计当中，这对整个剧院设计都产生了很大的影响。新的剧场必须拥有适应这些变化所需的舞台设施。

昆山市文化艺术中心是昆山市重点工程，整个工程无论是质量还是技术，都有极高的要求。我们的设计工作是为业主提供一套工艺先进、功能强大的系统方案，进而由舞台设备承包商用一流的技术、过硬的产品、优质的服务去实现该设计方案。为实现世界一流的设计，无论是总体设计还是极细小的局部，我们都提供了最新、最佳、最成熟的技术解决方案。处理好每一个细节问题，解决舞台设备系统的难题正是做好本项目的关键。

考虑到昆山市文化艺术中心将来运作的高要求，舞台设备系统是舞台安全高效运行的重要保障。我们设计所要求的计算机控制系统采用世界领先的科技，能充分发挥舞台设备的潜力、开拓各种演出的可能性，并提供了完善、合理、先进的舞台设备工艺设计方案，以保证现在和未来的高水平、高质量的演出。

新印象：设计实施的过程如何？有哪些值得分享的事件？

李跃文：整个设计过程是严谨的，也是一个充满挑战的过程。比如，主要用于昆曲演出的小剧场是多功能剧场，为了最灵活的使用方式以及充分利用场地的设计要求，我们为其配备了移动座椅以及平整地面上的灵活舞台。并配置了一定数量的双层升降台，这些升降台配备机电驱动系统，可以安全、快速、方便地转换布局。为使观众厅座椅灵活，并提供了一组座椅台车，可以根据每场演出的需要来调整这些座椅排列布局。如果不需要用座椅台车，座椅台车可以储存在升降台的下层台板上。设计最终效果可实现镜框式舞台、开敞式舞台、T台、延伸的T台、岛式舞台、宴会厅等多种布局形式。

在整个过程中，业主方负责人员的敬业精神和专业素养都给了我们很大的帮助，也给我们留下了深刻的印象。

新印象：这个项目的设计实践对今后同类项目的开展有哪些值得借鉴的地方？

李跃文：通过这个项目的设计，给我们今后设计类似的项目提供了宝贵的经验，也使得我们在今后的设计工作中再遇到此类剧场时能够更快、更清晰的结合当地文化，将当地文化融入到舞台设计当中。

New Impression: How do you feel about this project? What is its specialty?

Li Yuewen: In fact, the functions of every theatre are similar. For us, the primary work in the beginning of design is in-depth communication with the user (or investors, hereinafter referred to as the clients), understanding client's wish (using requirements of the theatre to be built) well, i.e., functional positioning. For the designers, it is very important to know the functional positioning of the theatre, since it is the basis and premise of theatre design. For Kunshan Culture and Art Center, it is an elegant and friendly palace of art. With good location and convenient transportation, it is an important part of Kunshan's city pattern. Designed by famous master architects in China and integrating the local cultural characteristics of Kunshan, the whole building is a perfect combination of local traditional culture and architectural art, as well as an architectural art treasure revealing local traditional culture everywhere.

New Impression: What are the difficulties and requirements on theatre design of Kunshan Theatre? Is there any case to be referenced? What are differences between the western theaters and this one?

Li Yuewen: Regarding stage design, it is difficulty, being similar to architectural design probably, is to combine modern stage technology with traditional art culture. In other words, it should not only implement the performance requirements of traditional Chinese theatrical pieces, but also meet various possibilities of traditional western theatrical pieces and modern popular operas as well as all future possibilities. What we should do is to understand the requirements of Chinese opera on stage and incorporate modern stage technology. For the owner and us, it is a multiple communication process as well as a learning process.

In recent years, theatre stage design has had great changes. The greater, finer and more exceptional performances call for more and better stage equipment than before. Thus, newer, higher and maturer techniques should be employed in the design of stage facilities and equipment, which has had significant impact on the design of whole theatre. New theatre should have stage equipment needed to adapt to these changes.

Kunshan Culture and Art Center is a major project in Kunshan. The whole project has high requirements on quality and technology. Our design efforts are to provide a powerful system scheme with advanced technology. In order to be a world-class design, we provide the newest, best and maturest technology solution to whether overall design or tiny parts. Dealing with the details and solving the problems of stage equipment system are the keys to finish this project.

With a view to the high operational requirements of Kunshan Culture and Art Center in the future and considering that stage equipment system is an important safeguard for the safe and efficient operation of stage, the computer control system required by our design adopts world-leading technologies which can give full play to the potential of stage equipment, explore the possibilities of various performances and provide perfect, reasonable and advanced process designing scheme of stage equipment to ensure the sophisticated and high-quality performances at present and in future.

New Impression: How did the process of construction work? Do you have any interesting storys to share with?

Li Yuewen: The whole design process is rigorous and challenging. For example, the little theaters for the performances of Kunqu operas are multi-functional theaters. For the most flexible way to use and the design requirement of making the most of fields, they are equipped with movable seats and flexible stages on flat ground. They are also configured with a certain number of double-layer lifting platforms equipped with electromechanical driving system which can shift layout safely, quickly and conveniently. For the flexibility of seats in the auditorium, a set of seat trolley is provided to adjust the arrangement and layout of seats according to the requirements of each performance. If the seat trolley is not required, it can be stored on the lower platen of the lifting platform. In the final effect of the design, proscenium stage, open stage, T stage, outstretched T stage, arena stage, banquet hall and other forms of layout can be achieved. In the whole process, the responsible personnel of the owner, with their professional ethics and qualities, help us greatly and impress us deeply.

New Impression: What can be used as the reference point on this design practice for future similar projects?

Li Yuewen: The design of this project provides valuable experience to our similar design projects in the future, and enables us to combine local culture faster and clearer and integrate local culture into stage design in the future project works when such theaters are confronted again.

金照彬，中天建设集团有限公司项目副经理

Jin Zhaobin, project manager assistant, Zhongtian Construction Group Co., Ltd.

Interview with Jin Zhaobin

Construction Difficulties and Solution Thereof

建造难点与应对·金照彬

新印象：这个项目在深基坑方面面临的问题有哪些？

金照彬：文化艺术中心工程基坑开挖面积大、维护结构长、开挖深度深是土建施工的难点之一。其中，基坑面积约25300 m^2，围护结构740延长米，基坑开挖深度1.05m~14.70m，创昆山之最。因中心基坑开挖深度较大，且场地土质含水率高，因此难度较大。在施工方案中对于其挡土系统、止水系统、支撑系统、栈桥系统等多方面进行了考虑，并实时监测。

新印象：施工中都是怎么应对的？

金照彬：首先是挡土系统，考虑到本基坑各侧环境的不同以及场地土体情况，本方案相应采取了不同的围护结构。其次，场地内浅层存在透水性较好的土体，且起伏较大，因此必须设置可靠的止水帷幕。根据场地土层分布情况，尤其是4层土的层顶标高分布情况，分别采用双轴搅拌桩和三轴搅拌桩进行止水：外侧挖深较浅的基坑采用双轴搅拌桩进行止水，中心基坑则采用受土性影响较小的三轴桩进行止水，且该工艺可施工深度较双轴搅拌桩大大增加。

而支撑结构设计是基坑围护设计的核心内容。对于支撑系统，为尽可能减小对周围环境的影响，需要一个强大的支撑系统；为了做到经济合理，又必须使用尽可能少的支撑材料；要挖土方便，减少施工周期，支撑之间还得留出很大的挖土空间，支撑结构设计面临众多矛盾。

通过上述分析，在中心基坑竖向设置一道支撑，支撑中心标高定为-7.10m。支撑材料选用C35钢筋混凝土，主支撑截面800×900mm，连杆截面尺寸600×600mm。圈梁采用C35混凝土，截面尺寸800×1200mm。另外，中心基坑东南侧由于受围护结构厚度限制，采用了直径相对较小的钻孔桩，这就需要强大的内支撑，因此在该区域设置了下琵琶撑。

另外，这个项目的栈桥系统也是根据施工需要而特意创建的，因中心基坑开挖深度较大，且出土不便，因此考虑设置混凝土栈桥作为中心基坑的挖土平台和出土通道。并使用基坑监测包括对环境的保护监测和对本围护体系的安全监测系统，预报施工过程中可能出现的问题。

新印象：对于这个项目而言，还有什么关键性的难点？

金照彬：艺术中心主体高支模是土建工程中另外一项难点。主体工程层高高，混凝土梁跨度大，混凝土梁悬臂长，主体钢结构和混凝土结构连在一起，采用大量预应力梁，内外结构均为不规则圆弧。主体整个造型为不规则弧线，弧线均为不规则形状，没有半径，主体的定位均以座标来定位。且一到五层均不相同。室内标高众多，每层楼面均有多个标高。所以根据以上情况，本工程选用了梁模板为18mm厚国产多层胶合板（915×1830mm）。所有模板全部在现场加工成型。以剧院3层为例（A-9轴至A-12轴交A-E轴至A-H轴），此区域从一层-1.53m楼面标高到四层楼面16.4m标高，高度达到17.93m，最大梁截面600×3300mm，跨度最大24m。

经过对提高昆山文化艺术中心曲面施工质量支模技术攻关团队的努力，在建筑施工方面，专门针对支模架的技术难点工程进行攻关，保证了主体结构的质量，达到了理想的效果，该技术攻关获得上海市建设工程QC成果金奖。

New Impression: What are the problems of this project in the deep foundation pit?

Jin Zhaobin: The foundation pit of the engineering of the Culture and Art Center needs to be excavated for a large area with long enclosure structure and deep excavation depth, which is one of the difficulties in civil construction, wherein the area of the foundation pit is about 25,300 square meters, the enclosure structure is 740 linear meters, and the foundation pit is excavated for 1.05 to 14.70 meters, which is top one in Kunshan. Because the foundation pit of the center is deeply excavated, and the water content of the site soil is high, the difficulty is severe. In the construction scheme, it has considered the earth retaining system, the sealing system, the support system, the trestle system, etc., and real-time monitoring is simultaneously carried out.

New Impression: How to solve them in construction?

Jin Zhaobin: Firstly, the earth retaining system, because of the variation of environments of all sides of the foundation pit and the condition of the earth body of the site, the solution correspondingly uses different enclosure structures, and secondly, there is the shallow earth body of which the water permeability is outstanding in the site, and the undulation is high, so reliable waterproof curtains must be arranged. According to the distribution of the soil layers of the site, particularly the distribution of elevations of the layer top of the 4th soil layer, biaxial agitation piles and triaxial agitation piles are respectively used for retaining water; and the foundation pit of which the excavation at the outer side is shallow uses the biaxial agitation piles for retaining water, the central foundation pit uses the triaxial agitation piles which cannot be highly influenced by soil for retaining water, and compared with the biaxial agitation piles, the construction depth of the technic is greatly increased.

The design of the support structure is the kernel of the enclosure design of the foundation pit. In terms of the support system, for minimizing the influence on the surrounding environment, it needs a strong support system. To be economical and reasonable, it must use the least support materials. To conveniently excavate, and shorten the construction period, a big excavation space needs to be left between supports, so there are many conflicts in the design of the support structure.

As analyzed above, a support is vertically arranged in the central foundation pit, and the elevation of the support center is set as -7.10m. C35 steel reinforced concrete is used as the support material. The cross section of the main support is 800×900mm, and the size of the cross section of the connection rod is 600×600mm. C35 concrete is used for the ring beam, and the size of the cross section is 800×1200mm. In addition, because the southeast side of the central foundation pit is limited by the thickness of the enclosure structure, the drilled piles of which the diameter is relatively small are used, so it needs strong inner support, and that is why Pipa supports are arranged in the area.

In addition, the trestle system of the project is specially created as required by construction. Because the excavation depth of the central foundation pit is high, and earth transportation is inconvenient, thus, the concrete trestle is arranged as the earth excavation platform and the earth transportation passage of the central foundation pit. The monitoring of foundation pit, including the systems for the monitoring on environment protection and the safety monitoring of the enclosure system, are used to forecast the possible problems in construction.

New Impression: Besides the above aspect, to this project, is there any other crucial difficulty?

Jin Zhaobin: The high formwork of the main body of the art center is another difficulty in the civil engineering. To the engineering of the main body, the floor height is high, the span of the concrete beam is big, the cantilever of the concrete beam is long, the steel structure of the main body and the concrete structure are connected together, a great quantity of prestressed beams are used, and both the inner structure and the outer structure are in an irregular circular arc shape.

The whole model of the main body is in an irregular arc shape, all arcs are irregular without radius, and coordinates are used for the positioning of the main body. It is varied from the 1st storey to the 5th storey. There are many indoor elevations. There are a plurality of elevations on each floor. Thus, according to above, the engineering adopts the domestic multi-layer plywood (915×1,830mm) of which the beam template is 18 mm thick. All templates are processed and shaped on site. Take the 3rd floor of the theatre as an example (A-9 axle to A-12 axle are crossed with A-E axle to A-H axle). To the area, from the floor elevation of -1.53m of the 1st floor to the floor elevation of 16.4m of the 4th floor, the height is 17.93m, the maximum cross section of the beam is 600×3,300mm and the maximum span is 24m.

With the effort of the team for improving the support template technology of the curve-surface construction quality of the Kunshan Culture and Art Center, to building construction, it focuses on conquering the technical difficulty of the support template rack so as to ensure the quality of the main body structure and get the ideal effect. The Golden Award of Shanghai Construction Engineering QC Achievement is awarded to the technological breakthrough.

范晓玲，昆山市住房和城乡建设局副局长，原昆山市城市建设投资发展有限公司副总经理\昆山文化艺术中心开发有限公司 总经理

Fan Xiaoling, deputy director of Kunshan Housing and Urban Construction Bureau, former vice general manager of Kunshan City Construction Investment & Development Co., Ltd. and general manager of Kunshan Culture and Arts Center Development Co., Ltd.

Interview with Fan Xiaoling

Challenging Cultural Buildings

具有挑战性的文化建筑·范晓玲

新印象：文艺中心方案一出台是否有感到一些设想将来实现起来会面临严峻的挑战与压力？值得国内外项目借鉴的地方有哪些？

范晓玲：为了了解项目，公司专门组织我们从国家大剧院开始，尤其对东南沿海的剧院考察，不仅是了解经验更是重点要了解建造中遇到的问题、存在的问题，结果发现，所有剧院建设中均遇到很多的困难，多多少少建成后均留有缺憾。因此越考察无形中的压力就越大。

这样的大空间，工艺和技术要求都不同于一般的项目，面临的挑战很多，如项目15m深的基坑在昆山都没有过，当时基地环境还不是很好，下面还有流沙。怎么把围护做好，怎么把安全做到位，可以说就是一个巨大的挑战。项目60%以上都是大空间，建筑的形态像一朵花一样，甩出去的部分很多。因此，钢结构起吊重量大和吊装距离远，在以往的项目也是没有的，施工期间我们采用了两台350、400吨的吊车。建筑师设计的钢结构有个飘带的造型，事实上它是一个三维的空间造型，需要由六根管子把它拼接起来，非常困难。负责钢结构施工的浙江精工钢结构有限公司的项目经理毛博士做了几十年的钢结构项目，他也说还很少碰到这么具有挑战的钢结构项目。

要说经验我个人觉得首先设计是关键，必须把问题在图纸阶段解决，让各种专业顾问同步参与。为了较好的完成文化艺术中心设计我们组织整个设计团队实地参观正在建设中、已完成的剧院，大家对工艺要求理解比较清楚后再上路。其次建立良好的沟通协调机制，使得各种问题高效解决、各种意见能充分得到提炼落实。再次就是谨慎选择好施工单位，设计得好，通过有能力的单位建设出来才是根本，文化艺术中心绝大数单位体现了一流的施工管理水平，使项目得以保质按期完成。

新印象：根据环境、地域的特点这个项目实施过程中面临过哪些问题？做了哪些大胆的尝试？在项目协调过程中，碰到过几次实施难点？过程如何？分别以什么方法应对的？

范晓玲：由于意识到项目的难度，觉得每一步都不容易，每一步都非常谨慎对待。项目的核心功能区是舞台，舞台基坑15m，地上高度32m，总共50m，是个类似筒的空间，对舞台设备安装来讲，施工难度很大。运行要求精度很高。为此，作为奥运会的舞台设备安装实施者——总装部，以他们的丰富独到的经验，严谨的军人作风严格而高精准地完成了设备安装。

大剧院外幕墙设计采用穿孔幕墙，不仅尺度大，高度高（最高的幕墙板高度达9.6m），而且总设计师崔愷院士要求要比较通透，不能有过多的结构来支撑，也正由于这样的难度先后有几家单位来参与，中途都被迫退出。最后我们请到了国内非常有经验的专家——来自上海的和甲幕墙设计咨询有限公司作顾问，研究多种

材料或工艺，最后利用飞机机翼生产工艺，巧妙地运用了飞机机翼的铆接工艺，不仅超薄、超长的铝板构造实现了以刚生柔的设计理念，并有效隐藏了支撑构件及泛光照明部件和方便检修的构件锁，实现了设计的目标。幕墙的玻璃采用超白玻璃、玻璃勒加隐形索的结构体系更增强了建筑的通透感，玻璃勒上还有图案，增强了细节处理的效果。

此外，设计者为追求室内部分素雅的效果，采用了GRG预制板，通过对国内建成的大剧院考察我们发现，这部分做得好则好，做不好会使整个建筑失分。为此我们专门考察了一批GRG生产厂家，与设计师一起研究，最后确保了工程的质量。

另外，多功能剧场是我个人所喜欢的一个装饰设计环节，设计师设计的三角形墙面可以翻动，使剧场墙面可以随意组合成任何一种需要的效果。

新印象：项目的一期部分现在基本完成，您认为对比方案，它的完成度如何？有哪些重大而有意义的改动？与您预想的感受有什么不同？

范晓玲：应该说基本上把设计师的理念体现出来了，并且更进一步完善了设计师的方案，比如在设计初期并没有这么多钢结构，部分原来采用混凝土的空间内装完成后层高低，通过建立三维模型，将一些结构改为钢结构，提升了空间高度，整个钢结构进行了重新设计、优化。虽然景观设计最后的实施与方案相比，对水的设计有所改动，但是也可以说是仁者见仁，智者见智。

新印象：建造过程中最让您印象深刻的或者最能打动您的是什么？

范晓玲：无论是设计单位、顾问单位，我们都有幸碰到一些技术精湛且有很高的职业道德的合作伙伴。通过不少极有见地的建议和提醒给了我很多支持与启发。项目是很多人无私的付出的成果，如崔愷院士虽然接手的项目很多，但每个月都抽时间来现场了解项目进展；再如总装部进行舞台机械安装的工作者两班倒，夜以继日的军人工作作风；在此，也感谢我们的公司建设团队的齐心协力，不计得失，忘我付出的每个日日夜夜。正由于各方一致的艰辛努力，试演出和两会才得以成功地在文化艺术中心召开。

New Impression: Did you feel huge challenges and pressures in realizing some parts of the project after the proposal was pointed out? What can be learned from similar projects both at home and abroad?

Fan Xiaoling: To understand the project, the Company specially organized us to investigate the National Centre for the Performing Arts, especially theaters in the southeastern coast. We shall learn not only experience but more problems in construction and existence. Finally, we found that all theaters had encountered many difficulties in construction and left more or less pities behind. So as the investigation went deeper, we were feeling more intangible pressure. Within such a large space, both technical and technological requirements were different from general projects, which would cause many challenges. For example, the 15m foundation pit had never been experienced in Kunshan with poor site environment and quicksand below. It is, as it were, a huge challenge to build the fence and ensure safety. Over 60% of the project was large space. Shaped like a flower, the building has a large part leaning away. As a result, the steel structure has large lifting weight and long hoisting distance, which did not exist with the past projects. During the construction, we adopted 350t and 400t cranes. The steel structure designed by the architect was shaped like a ribbon. In fact, it was a 3D spatial shape, which required six pipes to piece up. So it was very difficult. Dr. Mao, the Project Manager of Zhejiang Jinggong Steel Structure Co., Ltd., was in charge of steel structure construction with decades of experience in steel structure. He also said that such challenging steel projects were rare.

Speaking of experience, I think design is the key. The problems must be solved at the stage of drawings. All professional consultants must simultaneously involve in. To better complete design of the project, we organized the whole design team to visit the theaters under construction or completed. They did not set out until technique was clearly understood. Next, wonderful communication and coordination mechanism shall be well established to effectively solve all problems and fully distill all suggestions. Thirdly, we prudently chose constructors. Effective design combining with competent constructors is foundation. Majority of the participants in Kunshan Culture and Art Center showed first-class construction and management skills, which ensured completion of the project by high-quality and on time.

New Impression: What problems did this project meet during the construction in terms of the environmental and regional features? What attempts were made?

How many times did you meet the implementation difficulties during coordination process? What was the process going? What measures were adopted as response?

Fan Xiaoling: During the project, each step was difficult and challanging, and required our particular prudence. Core function area of the project was the stage with 15m of foundation pit and 32m above ground, 50m in total, forming a barrel-like space. For installation of stage facilities, construction difficulty is very large. Precision of operation requirement is very high. As a result, as installation constructor of Olympic Games stage facilities, Assembly Department completed facilities installation by rich experience, military style and high precision.

The theater employed perforated curtain wall in outer wall design. Besides large size and height (The highest curtain wall plate is as high as 9.6m), it shall be transparent and not supported by too many structures according to Cui Kai, the General Designer. Due to such a difficulty, many participants had to exit during the construction. Finally, we invited very experiential domestic experts as advisors from Shanghai DHD Curtain wall Design & Consulting Co., Ltd to try many materials or techniques. Finally, we adopted technique for plane wing making, i.e. riveting technique of plane wing. The ultra-thin and ultra-long aluminum plates have not only realized the design conception of flexibility from rigidity but effectively concealed supporting components, floodlighting components and component locks for repair convenience to meet the design objectives. Glass of the curtain wall adopts the structural system of ultra clear glass, glass ribs and invisible locks to enhance transparency of the building. The glass ribs have graphic patterns to enhance treatment of details.

In addition, for the interior, the designer has adopted GRG prefabricated plates for simple and elegant effect. According to the investigation on the grand theaters completed in China, this part may make the design more brilliant or damage effect of the building with poor construction. To this end, we specially investigated a number of GRG manufacturers and discussed with the designer to finally ensure quality of the Project.

In addition, multi-functional theater is a decoration design link I favor. The triangle curtain wall designed by the designer can be turned to make walls of the Theater freely structured into any effect needed.

New Impression: Phase I of the project has been largely completed. What do you think about its completeness comparing with its original proposal? What are major and significant changes? What are the differences from your expected effect?

Fan Xiaoling: As it were, it has demonstrated the design conception of the designer and further improved proposal of the designer. For example, at beginning of the design there were no so many steel structures. For some space originally adopting concrete, the storey height would be low after inside finish. Through 3D modeling, some structures were changed into steel structure to raise spatial height. The whole steel structure has been redesigned and optimized. Though design about water was changed to some extent in the final implementation of the landscape design as compared with the proposal, it is just different options.

New Impression: What did impress or move you most during the process?

Fan Xiaoling: It is very lucky that we have had some partners, both designers and advisors, with excellent skills and high professional ethics. I have received much support and inspiration from many insightful suggestion and advice. The project is the result of much selfless contribution. Though Cui Kai undertook many projects, he went to the site each month for information on project progress. Workers from Assembly Department worked on stage equipments installation by two shifts. They demonstrated military working style around the clock. Now, I appreciate the disinterested joint efforts of construction team of the Company during the days and nights of selfless contribution. On basis of the tremendous efforts from all parties, trial performance and People's Congress & Political Consultative Conference were successfully held in the Kunshan Culture and Art Center.

第三章 Chapter 3

文化·传承·思考
Culture·Inheritance·Thinking

本章悉心梳理了昆山市厚重的人文历史、浓郁的地域特色、经典的文化传承等传统脉络，揭示了建筑师创意思维的由来；并通过采访来展现昆山的城市规划蓝图，勾勒出昆山规划与城市设计的思索之旅及其与建筑个体的相互关系。

This chapter carefully sorts out Kunshan's profound culture and history, strong geographical characteristics and classic cultural development and explores the origin of architects' creative ideas; and through interviews unfolds the Kunshan urban planning blueprint and depicts the relationship between the journey of Kunshan planning and urban design and the individual buildings.

采访：许振敏，曹勇 Interview with Xu Zhenmin, Cao Yong

理念·功能·剖析
Idea·Function·Analysis

为实现“房在园中”的构思意图、建筑造型特点及功能要求，甲方、设计、材料、施工各环节密切合作，在制作模型、实验、模拟，反复测试与论证的基础上，最终不仅完整地实现目标，并为业界提供了很多值得借鉴的设计与实践经验。

To achieve the design idea, architectural features and functional requirements of "room in the garden", architects worked closely with clients in terms of design, materials, construction and other aspects. On the basis of modeling, experiments, simulations, repeated testing and demonstration, they finally achieved the whole design objective, providing the industry with a lot of design and practical experience.

采访与文章：崔愷，阿兰·博尼，李跃文，金照彬，范晓玲 Interview & essay: Cui Kai, Alain Bony, Li Yuewen, Jin Zhaobin, Fan Xiaoling

城市·设计·生活
Urbanization·Design·Life

昆山规划以及文化艺术中心的建造，汲取了国外经典案例的规划思想。不仅给居民带来了福祉，并为昆曲等传统文化的再生提供了一个互动舞台，让建筑自然植入城市，选择了一条适合城市与其居住者们的道路。

Kunshan planning and the construction of Kunshan Culture and Art Center draw on the planning ideas of foreign classic cases, providing not only benefits to the residents but also a platform to Kunqu opera and other cultures to revive and interact with people, implanting buildings naturally into the city with a path suitable for both the city and its residents.

采访：周继春，姚伟宏，韩明 Interview with Zhou Jichun, Yao Weihong, Han Ming

3.Urbanization·Design·Life

城市·设计·生活

进入20世纪以来，欧美中等城市如巴塞罗那，曼彻斯特，布法罗等，在不同时期面临城市转型的契机，其规划和城市设计方案先后意识到公共空间的重要意义，都是以审慎的态度把公共空间放在第一位来考虑。

这些城市均有效而颇具技巧地处理其中的文化建筑和城市肌理的关系。如今，巴塞罗那等城市的改造已成为世界城市规划的典范。建筑大师埃罗·沙里宁曾说过："建筑就像一本打开的书，从中你能看到一座城市的抱负。"今天的昆山规划尤其是文化艺术中心的建造，无疑汲取了以上这些城市的精神特质与符合人文精神的规划思想，展示出其向现代化城市转型的有效方式与计划。从本章收录的文化艺术中心建成后的演出剧目、市民反馈、国外经典城市空间设计案例和相关访谈等内容中可以感受到，在同样实现文化引领城市发展的目标过程中，文化艺术中心不仅给广大居民带来了更多层面的精神文化福祉；并为昆曲等传统文化涅槃再生提供了一个与各阶层市民互动的舞台；同时，在昆山面临城市转型的时间节点，文化艺术中心无疑把握住了契机，让建筑自然地植入城市，更有机地联接既有的城市肌理，在规划与建筑设计实施过程中，选择了一条符合回归人文精神的国际设计潮流并适合城市本身和其居住使用者的道路。

Entering 21th century, European and American medium-sized cities such as Barcelona, Manchester, Buffalo were facing good opportunities of transformation in different periods, whose urban planning and designs had realized the importance of public space and put it in the first consideration in a prudent manner.

These cities effectively and skillfully handled the relationship between cultural architecture and urban texture. Today, the transformation of Barcelona and other cities has become the model of urban planning in the world. Architect master Eero Saarinen once said, "Architecture is like an open book, from which you can see a vision of the city." Today's Kunshan planning, especially the construction of Kunshan Culture and Art Center, has no doubt learned from the characteristics and humanism planning ideas of these cities, demonstrating its effective way and plan to shifting to a modern city. From the repertoire, public feedbacks, foreign classic urban space design cases and related interviews and other contents collected after the completion of the center in this chapter, we can appreciate that in the objective of culture leading urban development, Kunshan Culture and Art Center not only brings more spiritual and cultural entertainment to the general residents but also provide a good stage for Kunqu opera and other traditional cultures to revive and interact with people from all walks of life. Meanwhile, at this juncture of Kunshan urban transformation, the Culture and Art Center will undoubtedly seize the opportunity to naturally integrate buildings into the the city and more closely connect those buildings with existing architectures. In the process of planning and architectural design, we choose a path that meets with the international humanism design trend and fits the city and its residents.

NATIONAL
FOOTBALL
MUSEUM
HISTORY
SKILL
ART
FAITH
PASSION
FOOTBALL

SHEA'S
BUFFALO
SHEA'S
PERFORMING ARTS CENTER
Wegmans

2012年12月20日，昆曲折子戏《三岔口》剧照
On December 20, 2012, the Kunqu opera SanChaKou was on show.

2012年12月26日，大型服装舞剧《金舞银饰》剧照
On December 26, 2012, large dancing epic Gold Silver Dance Performance was on show.

2013年1月20日，话剧《恋爱的犀牛》剧照
January 20, 2013, drama Rhinoceros in Love

2013年1月13日-14日，音乐剧《宝莱坞明星》剧照
January 13-14, 2013, musical show Bollywood Stars

2013年2月5日，杂技剧《天鹅湖》剧照
February 5, 2013, drama Swan Lake

2013年3月2日，《杨坤1-32全国巡回演唱会》剧照。
March 2, 2013, "Yang Kun 1-32 national concert tour.

2013年3月8日，《中央歌剧院中外歌剧歌曲合唱音乐会》剧照。
March 8, 2013, Famous Chinese & Foreign Opera Songs Concert by China National Opera House.

2013年3月18日，话剧《分手大师》剧照。
March 18, 2013, drama Break-up Master.

2013年4月23日《东方三绅士男高音组合2013全国巡演昆山站》剧照。
April 23, 2013, national tour Three Oriental Gentleman 2013 in Kunshan.

2013年5月25日,《功夫传奇》剧照。
May 25, 2013, Kung Fu Legend.

2013年5月31日, 话剧《风华绝代》剧照。
May 31, 2013, drama The Heyday.

2013年7月30日, 大型歌舞诗剧《四美图》剧照。
July 30, 2013, large song and dance drama Four Beauties.

2012年11月3日，市民对新建成的剧院充满期待，办会员卡的市民排起了长队。
On November 3, 2012, the residents were full of expectation to the newly built theatre and kept a long queue to register the membership card.

2012年11月3日昆山文化艺术中心开票典礼举行。未来的演出昆山文化艺术中心的工作人员正为踊跃办会员卡的市民们登记。
On November 3, 2012, the Ticket Box's opening ceremony of Kunshan Culture and Art Center has been held. The future staff of center registered the citizen who wanted to get a membership card.

2013年1月10日，昆山文化艺术中心落成典礼中的观众。
On January 10, 2013, the founding ceremony of Kunshan Culture and Art Center.

2013年1月3日，昆山市人民代表大会第十六次代表大会第二次会议的会议现场。
On January 3, 2013, on the conference spot of the Kunshan people's Congress of the second meeting of the Sixteenth Congress.

2013年1月14日，联合昆山市义工联合会组织柏庐社区的15位老人观看演出。
On January 14, 2013, the center associated with the Kunshan's Social Work Association invited 15 old men of Bolu community to watch the program.

2013年1月23日，昆山文化艺术中心多功能厅隆重举行了"爱在阳光下"——2013年保利携手"中大未来城"公益行，昆山市民音乐会暨新春原创互动游戏儿童剧《超级玛丽》新闻发布会。

On January 23, 2013, in the Multi-function Hall of Kunshan Culture Art Gallery, "Love in Sunshine" has been celebrated—the 2013 non-profit tour held by Baoli and "Zhongda Future City", the Kunshan Citizen. Concert and the press conference of New Year original interactive game child play Super Mario. Show in Kunshan".

2013年2月14日（大年初五），来自上海鼎川文化传播有限公司的儿童剧演员们在昆山文化艺术中心的多功能厅为昆山小朋友们带来了两场原创互动游戏儿童剧《超级玛丽》。

On February 14, 2013, the fifth day of traditional Chinese new year, the child play crew of Dingchuan Culture Broadcasting Co., Ltd. came from Shanghai, brought two original interactive game child play Super Mario for kids in Kunshan in the Multi-functional Theater.

2013年2月15日下午14:00，昆山文化艺术中心多功能厅迎来了首次文化艺术讲堂活动，由京剧名家尚长荣老师为观众带来精彩的专题讲座《激活传统，融入时代：我的艺术人生》。

At 2 o'clock p.m. on February 15, 2013, the Multi-functional Theater of Kunshan Culture and Art Center welcomed the lecture on culture and art for the first time; Shang Changrong, a master of Peking Opera gave audience a fabulous lecture on the topic of Activating Tradition, Merging in Times: My Life about Art.

2013年5月11日至19日期间，免费向公
展"举行。著名的圆明园回归国宝——
及5件青铜器、40余件玉器正式与市
From May 11 to 19, 2013, "Poly Chi
Palace Tour Show in Kunshan" has
famous returning treasure of Sum
statue Ox Head, Tiger Head, Mon
bronze vessels, 40 jade objects face

2013年5月11日，
的观众。
On May 11, 201
Show in Kunsh

评论与反馈 Comments & Feedback

从昆山文化艺术中心修建到投入使用，其间一直备受昆山网友们的关注，他们成为除业主和设计者之外，见证和督促这个文化艺术中心成长、完善的第三种力量。

During the period from construction to use, Kunshan Culture and Art Center has drawn a great number of attention from netizens, becoming the third force to witness and urge the development of center except the owners and designers.

@inver:昆山文化艺术中心现在建到什么程度了？斑竹阿能发点现场图片上来？是不是已经基本完全好了？
2012-10-22 16:24 昆山论坛
@inver: What is the construction progress of Kunshan Culture and Art Center now? A-neng, would you please post some pictures? Is it roughly completed?
2012-10-22 16:24 Kunshan BBS

@天乐:真心希望政府加大对文化产业的投入。扶持更多的小微文化企业。希望中心有更多的图片展。微电影作品展。全民参与。
2012-10-31 19:36 昆山论坛
@Tianle: Sincerely wish the Government will increase input to culture industry to give more support to mini/small culture enterprises and hold more pictures and micro films works exhibitions, forming full participation by all people.
2012-10-31 19:36 Kunshan BBS

@ginvo:今天也去了，看到保利院线招牌了，冲着保利二字很期待。（如果能）详细告知开业计划和影院设备。一定前去捧场！！！！！
2012-11-03 21:34 昆山论坛
@ginvo: I visited it today and saw a poster of Poly Theater. It deserves to be expected just for the word "Poly".(If possible,) please announce commencement plan and theater facilities in detail. I must go to visit the opening ceremong!!!!!
2012-11-03 21:34 Kunshan BBS

@小兔宝宝:城西发展的真好,作为新昆山人，感觉挺幸福的呢。
2012-11-27 21:25 昆山论坛
@Xiaotubaobao: The development is rapid in the west of the City. As a person in New Kunshan, I feel very happy.
Published on: 2012-11-27 21:25 Kunshan BBS

@幽芯梦:昆山早该有如此陶冶情操的地方了！
2012-11-29 23:04 昆山论坛
@Youxinmeng: Kunshan needs such a nourishing place!
2012-11-29 23:04 Kunshan BBS

@acttqq:昨晚在昆山文化艺术中心看了曹云金的《分手大师》的狠(很)不错，现场笑声不断，我是第一次看话剧舞台剧，真的很好看，我把肚子都笑痛了！位置坐得还可以，前几排去的人也很多啊，楼下基本都坐满了！随意拍了几张照片。
2013-03-19 新浪微博
@acttqq: Last night, I watched Break-up Master performed by Cao Yunjin at Kunshan Culture and Art Center. It was wonderful and we laughed all the time. It was the first time for me to watch a theater play. It was really great. I burst my sides with laughter! I got a favorable seat. The first rows and the seats on the ground floor were largely filled! I also took some photos.
2013-03-19 Sina Weibo

@朗朗莫离:终于排队办到两张卡！我们大约九点十五到，快到队尾了。第一个办卡的竟然6点就过来排队了，太强了。现在排队选票中……
Android手机客户端2012-11-03 21:25
@Lanbglangruoli: Finally, I got two cards after long queuing! We arrived there at about 9:15 and stood at end of the queue. The first person of the queue actually arrived at 6:00. He was really something. Now, I am queuing for choosing tickets…
From Android Mobile App
2012-11-03 21:25

@离线泱泱:之前去过一次，各方面都是不错的 。
2013-04-24 08:59 昆山论坛
@Lixianyangyang: I visited once. All aspects are quite fine.
2013-04-24 08:59 Kunshan BBS

@鹿城瑞海阳光：很值得一看，再回首一起的日子。粉丝超赞，不老的歌迷 。(回复5月9日 " 我和我的朋友"姜育恒演唱会。)
2013-5- 9 23:04 新浪微博
@Luchengruihaiyangguang: It deserves watching and I can't help remembering the past time. Fans are great, and the song enthusiasts immortal. (Reply to May 9 "My Friends and me" Jiang Yuheng Concert.)
2013-5- 9 23:04 Sina Weibo

@宇宙大魔王啊鲁：谢谢，给我们一次免费欣赏昆曲的机会。（回复@昆山文化艺术中心:国宝展惠民大放送——免费昆剧折子戏）
2013-5-15 08:51 新浪微博
@Yuzhoudamowangalu: Thanks. It gives us a free chance to enjoy Kunqu opera (Reply to @ Kunshan Culture and Art Center: National Treasure – Kunqu Opera Highlights)
2013-5-15 08:51 Sina Weibo

@北京西路64:第一次近距离接触不能算是国宝的"国宝"——圆明园四大兽首：牛、虎、猴、猪。刚看，做工细腻，形体精致，古铜苍沧，美。
2013-5-13 11:36 新浪微博
@Beijingxilu64: It has been the first time for me to enjoy "National Treasures", i.e. the four major beast heads from the Old Summer Palace, Ox, Tiger, Monkey and Pig. At the first sight, they are made of bronze with delicate and fine workmanship. So they are very beautiful.
2013-5-13 11:36 Sina Weibo

@爱摄影Me浩然蒸气:不用去北京都能亲眼目睹国家文物的真容，而且还是免费观看展览，于是乎下午挎着GF2就去了，路上随拍了些照片。
2013-5-14 19:11 360浏览器超速版新浪微博
@Aisheying Me haoranzhengqi: You can personally see the true appearance of national cultural relics and enjoy the exhibition for free without going to Beijing. As a result, I went there with my GF2 camera and took some photos along the road.
2013-5-14 19:11 From 360 Viewer Superspeed Version Sina Weibo

@昆保皮小蓟:其实我之前也没看过现场版的，在电脑上看过，说实话真无感。但是在现场看过之后才发现啊，真的跟电脑上看很不一样啊，现场版的会有很多环境因素吸引观众，让观众更好地理解剧情，融入到演出中。
2012-12-04 09：44
@Kubaopixiaopu: In fact, I had never watched on site except on TV. So it was short of real experience. After visiting the site, I felt it is really different from that on the computer. Real building attracts more audience by many environmental factors and help the audience to understand the play and merge into the performance.
Published on: 2012-12-04 09：44

@mumar:今天看看后续工作快完工了，终于可以一睹芳容了啊。蛮大气的啊，开业了就去 。
2012-12-27 19:55
@mumar: During visit today, I found it would be completed soon. Finally, I will see its charming appearance. It is really magnificent. I will go there as soon as it is opened to operation. 2012-12-27 19:55

@害羞的苹果：感觉不错，完工了去看看支持一下。 2012-10-15 18：36
@Haixiudepingguo: It looks good. I will visit it after its completion to express my support. 2012-10-15 18：36

@chesterzhu:文化艺术中心要开IMAX？以后看电影不用去市区了。
2012-12-12 20：33
@chesterzhu: Kunshan Culture and Art Center will introduce IMAX? In future, we don't have to visit the urban area for films.
2012-12-12 20:33

@fong:没想到昆山的文化环境这么好，福气啊 。
2012-11-10 昆山论坛
@fong: It is never imagined that cultural environment of Kunshan is so wonderful. It is really a blessing.
2012-11-10 Kunshan BBS

@远远的云儿：《恋爱的犀牛》吗？我们排队买了3张，值得。2012-11-05:23:11
@Yuanyuandeyun ：Rhinoceros in Love? We have bought 3 tickets by queuing. It is worthwhile. Published on: 2012-11-05:23:11

@dlshmily9：音乐剧《宝莱坞明星》，所到之处票房一抢而空，牛的啦！！
2012-12-25 12:33昆山论坛
@dlshmily9：The tickets for the musical drama Bollywood Star are sold out everywhere. It is really cool!!
2012-12-25 12:33 Kunshan BBS

@懿涵 ：让我们相约艺术中心，1.21号的《大河之舞》。
2012-12-20 21:55
@Yihan ：Let's gather in the center for River Dance on January 21.
2012-12-20 21:55

@圣诞结：文化艺术中心还没去过呢，我想应该值得去看一看的。
2012-11-21 20:44昆山论坛
@Shengdanjie: I have not visited Kunshan Culture and Art Center yet. The place is worth a visit.
2012-11-21 20:44 Kunshan BBS

Essay by Audiences

Feelings and Comments

观众观赏感受与评论

和艺术结缘，应该是从大学开始的。那时学校有很多社团，爱影人，爱乐社，光影社。我参加了爱影人，第一次活动，指导我们的是一个看上去50多岁实际上已70多岁的于老师，样子很和蔼。他给我们认真分析着电影的情节，画面，艺术成就。当时就被深深吸引了，原来电影音乐和戏剧还有这么多名堂，这么大的魅力啊！

日渐成型的设计

记得前年的某一天，在网上看到一则新闻，说昆山将要建设文化艺术中心，特别激动。后来一直关注着工程进度，在网上搜一些设计图，设计理念等资料。公司班车经过前进路，每天都能看到艺术中心一天天的变化。总算，盼到了开业的这一天。2012年底的一个周末，我来到艺术中心，似乎刚刚完工，工人们还在进行最后的收尾工作。艺术中心看上去很气派，很有艺术范儿。绕着走了一圈，看到设计图纸变成现实真的很让人感慨。走到售票厅，装修很精致。询问了工作人员，都很礼貌，很有耐心。我毫不犹豫办了会员卡。

建成后的实地体验

第一次看的演出的是《宝莱坞明星》。约上妹妹，也让她感受一下。第一次在家门口能看到充满印度风情的音乐剧真的很激动。华丽的服装，绚丽的灯光，激情的舞蹈，民族情调，美妙的音乐。走出剧院，刚才的那些情境像做了一场梦一样。回到现实，会觉得不适应了。这是真的，刚才真的是看的音乐剧吗？《恋爱的犀牛》是我第一次看话剧，纠结的剧情，激情的表演，偏执的爱情，让我想了很多。不愧是享誉全国的先锋话剧，不愧是年轻人的爱情圣经！现场感受戏剧和看电视剧完全是不同的感受，你和演员同悲伤，共欢笑！昨晚下雨了，很冷，吃完饭，骑上车，披着雨衣，气喘吁吁地赶到大剧院。《大河之舞》这场是人最多的，当灯光暗下，音乐响起，演员们跳着动感的踢踏舞，观众们沸腾了。对，这就是我一直梦寐以求的踢踏舞，世界上最经典的踢踏舞，我想大声告诉于老师，我看到了！旁边的小朋友也不由得跟着跳起来，真可爱！真心谢谢昆山文化艺术中心，让我们普通老百姓在家门口就能欣赏到最经典的艺术作品。

—

文：杨振华

Forming ties with art should start when I was at university. At that time, there are many clubs at campus, such as Film Lovers, Music Club, Lighting Club, etc. I joined in Film Lovers; Mr. Yu, the teacher who guided us at the first event, was more than 70 years and he looked very nice, although he looked like 50 years old. He gave us a careful analysis of the film' s plot, images and artistic achievements. At that time, I was deeply attracted, knowing that film music and drama have so many tricks and how fantastic they are!

Increasingly Shaped Design

One day of the year before last year, I remembered that I found a piece of news saying that a Culture and Art Center would be built in Kunshan. I was so excited then. In the following days, I had been paying attention to the progress of the project, and searched some design drawings, design concepts and other information online. As the scheduled bus of our company passes Qianjin Road, I could see the Art Center become taller and taller every day and reveal the true appearance. Finally, the opening day came. In a week at the end of 2012, I came to the Culture and Art Center, and it seemed to be just completed; workers were still in the final finishing work. The garden workers were still planting rows of bamboo. The Culture and Art Center looked very impressive and artistic. Taking a detour around, it really provoked my profound emotion to see the design drawings become a real building. Walking into the ticket hall, I found that the decoration is very delicate. I enquired about information from the staff and they were all very polite and very patient. I did not hesitate to apply for a membership card

On-the-spot Experience after the Completion

The first show I watched was Bollywood Stars. In invited my sister in advance and made her experience too. It was really excited to watch musicals full of Indian style at doorstep. There were gorgeous costumes, brilliant lighting, passionate dance, ethnic ambience and wonderful music. Walking out of the theater, the circumstances were just like a dream. Back to reality, I was like a fish out of water. What I watched just now was really a musical? The first modern drama I watched was Rhinoceros in Love; the tangled plot, passionate performances and paranoid love reminded me of a lot of things. It really deserves to be called a nationally renowned modern drama and a Love Bible for young people! Field experience at theater and watching TV at home are completely different, because you feel sad and laugh with the actors together! It rained last night and it was quite cold; after dinner, I rode my bike and dressed in raincoat, and then rushed breathlessly to the Grand Theatre. Riverdance, this show had the most audience; when the lights got dim, the music sounded, and the actors performed dynamic tap dancing, the audience' s blood boiled. Yes, this is the tap dance and the most classic tap dance in the world that I always dream of, and I want to tell teacher Yu loudly, I see it! The children around could not help dancing. How cute they are! I want to thank Kunshan Culture and ArtCenter sincerely, because it enables ordinary people to enjoy the most classic works of art at doorstep. I look forward to more brilliant performances in the future!

—

Text: Yang Zhenhua

记得初高中那会儿，我与同学一道在苏邮淘碟，摩登天空、口袋音乐等独立音乐厂牌，这是我对文艺圈的第一认识。而后在上海求学六年，从独立音乐出发，去了不少演出现场，前阵子听说昆山有了自己的文化艺术中心，作为长期的文艺爱好者，很是兴奋，恰巧有朋友赠我一月十三日晚《宝莱坞明星》的票，出于对印度风情的好奇，我与友欣然前往，开启了与昆山文化艺术中心的缘分之旅。

交通与室内环境

由于第一次去怕不认识路，我们比演出时间早了近一小时就到了。体育场边上，地理位置相对还是比较方便的，对面就是车站，还有夜宵线，晚上结束可以搭公交回去，这点让我们倍感贴心。在室内转了几转，旋转楼梯、不规则的楼层等，都是美观又现代的设计，在这样的地方享受文艺饕餮，着实享受。室内还设计有咖啡厅的休闲场所，对等待演出的观众也是贴心之举。

完美的设施体验与演出

与上海的东方艺术中心相比，昆山文化中心的音效、布景等硬件设施都在一个水平，观众可以专心致志欣赏一台美妙的演出。值得一提的是环形观众席的设计，我是看台票，去的比较早，就试了下各个角度的位置，基本每个位置都能毫不费力地清晰看清楚舞台的盛况，为所有观众都尽量考虑了最好的视野。以往看的舞台剧，日本的居多，压抑的悲剧色彩浓重，很少看《宝莱坞明星》这样绚烂无比激情四射的风格，真是大开眼界。当女主角陷入梦想与现实的纠结中，我仿佛看到了过去在我身上的桎梏，我的思绪随着她的身体起舞，回到了那个激荡的年代。在女主角的爷爷去世那个篇章，荧蓝背景下的独舞，全场唯有清脆的铃铛声，我几欲落泪。好的作品无关国别，舞者诉说的是印度文化的变迁，我想到的是我周遭岁月的翩跹。色彩斑斓的印度风情果然名不虚传，热情得就像冬天里的一把火，把这么一部励志、怀旧的故事舞到了我心头，暖意滋生，回味无穷。

—

文：林丸丸子

I remembered that I bought discs at Jiangsu Postal, such as modern sky, pocket music and other independent music labels, at the junior and senior school days, and this is my first knowledge of literary circles. Then I studied in Shanghai for six years. Starting from the independent music, I went to a lot of live performances, and I began to watch modern drama and stage play under the influence of my friends. A while back I heard that Kunshan opened its own Cultural and Arts Center, I was quite excited as a standing literature and art lover. A friend happened to give me a ticket of Bollywood Star on January 13; being curious about Indian style, I am pleased to go with my friend, which started a trip of karma with Kunshan Cultural and Arts Center.

Transport and Indoor Environment

As it' s the first time for us to go there and we were afraid of not knowing the way, we arrived there nearly an hour earlier than the performance time. Besides the stadium, the location is still relatively convenient; the station is just on the opposite, and there is also a night line, so we can go back home by bus when the performance is over, which makes us feel more intimate.

Making a few turns in indoors, both spiral staircase and irregular floors are beautiful and modern design, so it' s a real treat to enjoy literature and art feast in at such a place. There are also recreational areas such as cafe, and it is also an intimate act for the audience who wait for the show.

Perfect Facilities Experiencing and Performances

Compared with the Shanghai Oriental Art Center, the sound, scenery and other hardware facilities are all at the same level, so the audience can be thoroughly engrossed in enjoying a wonderful performance. What is worth mentioning is the design of circular auditorium; having grandstand tickets, I went there earlier and tried the positions of every angle; the audience can watch the grand occasion on stage at each position, and it considers the best field of view for all the audience.

Most of the stage plays I watched are Japanese ones and they all had a strong depressed sense of the tragedy; I rarely watch the stage plays of so very gorgeous and passionate style like Bollywood Stars, so watching it is really eye-opening. When the heroine was caught in the tangle between dreams and reality, I seemed to see my past shackles, and my thoughts danced with her body, coming back to the era of agitation. When it came to the chapter in which the heroine' s grandfather died, the solo dance under fluorescent blue background and the only crisp sound of the bells made me almost shed tears several times. Good works are independent of countries; although the dancers expressed the change of Indian culture, I was thinking of my past years. Colorful Indian style really deserved the reputation, and it was as warm as a fire in the winter; it danced such an inspirational and nostalgic story at my heart, breeding warmth and leading me to endless aftertastes.

—

Text: Lin Wanwanzi

巴塞罗那的公共空间

Public Space in Barcelona

巴塞罗那把公共空间对城市的影响放在第一位的设计思想，
使这座城市易于阅读，具有弹性。
The design idea of prioritizing the impact of public space on Barcelona makes the city ever more accessible and resilient.

Text
张道铭

在世界各地的城市之中，巴塞罗那长久以来都以高质量的建筑与都市公共空间引以为傲。即使西班牙正面临着经济困境，这个城市依然是世界建筑朝圣者最重要的目的地之一。都市公共建设重新整理了城市的空间架构，也改变了市民的生活形态。细致处理的公共空间，也让它成为欧洲城市中公共空间设计的翘楚。

质与量兼俱的公共空间规划建立在扩张计划的架构上，这是由工程师Ildefons Cerdà于1855年提出的，当时的城市在人口遽增的状况下产生诸多公共卫生的问题。Eixample为一个格子街廓系统。街廓成八边型，中间则留出住户共享的内院。道路设计非常宽敞，最窄处仍有20m。Cerdà的街廓设计并没有被完整执行，街廓内的小广场最后被住户的增建完全填满。即使如此，他的规划仍然让都市在快速扩张的同时，考虑到公共空间与光线需求，保障了居民的生活质量。

公共空间被视为一个基础设施。公共领域的建设不是单纯以美化为出发点，而是借此再造城市中被遗忘的地区，改善复杂的社会问题。1979年的首度民主普选让公众的利益开始受到重视，再加上有了连续三位同属左派的市长大力支持让政策持续进行，创造了今日的巴塞罗那。

巴塞罗那的公共空间营造可大致分成三个阶段 。第一阶段始自20世纪80年代，市政府尝试藉由公共空间再造与小型公共设施，让城市中一些几乎被遗忘的衰败区域重新恢复活力，最典型的例子便是旧城区的改造。这个实验性的政策获得了最初的成功。第二阶段是奥运准备期，整个都会区的大众运输与主要公共设施都在此时兴建并串联成一整体。第三个阶段介于奥运之后直到2010年，城市变得更有企图心，公共建设向南与北的郊区延伸兴建大型基础设施，如国际机场与商港。

巴塞罗那将奥运的建设区域整合入自1980年代开始的城市规划架构中。共分成四个区域，第一个区域为奥运主要运动馆群，座落于西南方的Montjuic山丘上。让Montjuic成为机能完善的都市公园。第二个区域是水岸的奥运村，水岸空间与港口的重新整顿让市民重新得以接近海洋。第三个区域位于西北方的Vall d’Herbron，借此重新整理这个都市化失序发展的郊区。第四个区域则位于市区Diagonal，重建既有的公共与运动设施。市政府利用新的公共设施来创造新中心，催化街区的活力。奥运的建设不只是为了国际赛事功能上的需求，也对当时的都市问题诸如衰败、弃置、都市漫延等找出解

开篇：水岸步行空间介于奥运村与海岸城区Barceloneta 之间。

对页上图：西班牙广场Plaza de Espanya，从Richard Rogers 重新设计的购物中心New Arenas Center远眺。1929年的世界博览会就在此举办，现在转变为巴塞罗那展览中心。远方山丘即为奥运主场馆的所在地蒙特胡依克山，造型独特的电视塔为著名的建筑师卡拉特拉瓦的作品。

对页下图：巴塞罗那当代美术馆，立面映照出周围的历史建筑，馆前广场成为年轻人的聚集处。

本页，从上至下：旧城区空间再造：Raval区巴塞罗那当代美术馆，由Richard Meier & Partners设计。当代美术馆前面的广场。当代文化中心，由HelioPiñón and Albert Viaplana设计。馆前的广场可举办许多户外表演。

Opening page: Passeig Maritim de la Barceloneta (waterfront walking space) is located between the Vila Olimpica (Olympic Village) and the waterfront district Barceloneta (photo: Ilaria Mancini) .

Opposite page, above: Plaza de Espanya, viewed from the New Arenas Center redesigned by Richard Rogers, where the 1929 World Expo took place. It is now the Barcelona Expo Center. The massif afar is Montjuic, where the main venue for the Olympic Games was located. The specially shaped TV tower was designed by famous architect Calatrava (photo: Ilaria Mancini) .

Opposite page below: Museu d'Art Contemporani de Barcelona (MACBA)has the surrounding historic buildings reflected in its façade glass. The front square becomes a hot hang-out spot for the young people (photo:Zhang Daoming) .

This page, from top to bottom: Remaking space in old districts: MACAB in Raval District, designed by Richard Meier & Partners (photo: Ilaria Mancini); PlaçadelsÀngels plaza right in front of (MACBA) in Raval District (photo: Ilaria Mancini); the Centre de Cultura Contemporània de Barcelona was designed by HelioPiñón and Albert Viaplana. The front square is available for various outdoor shows (photo: Zhang Daoming)

决的策略。

奥运的主场馆1929年巴塞罗那举办世界博览会时就已落成，但多年之后逐渐被遗忘与弃置，直到为了申办奥运，它才再度被注意到。赢得竞图的是由Correa, Milà, Margarit, Buixadé,Gregoretti 组成的国际规划团队。新古典式的建筑立面被完整保留，但是主要结构跟容量全都重新设计，成为可容纳五万六千人的奥运体育馆。Calatrava也为主场馆前的广场设计新的地标。值得一提的还有奥运跳水池，原有的游泳池为了奥运重新变更设计，为建筑师Antoni de Moragas的作品，地景结合的设计让观众与运动员都能享受俯瞰巴塞罗那的美景。整个Montjuic的公共空间被重新规划，破旧的建筑物获得新生，公共空间的可到达性增加，藉由巴士与览车可便捷地到达山顶，山丘上的都市开发受到控制，保留大量的绿地，在这个阶段真正成为可被市民享用的公园。

沿着城市的Baixada(斜坡之意)从北走到南，便可体验到连续整体的公共空间及其互动关系。首先看到的是Gracia 区的路口广场，广场所在的Gracia街区尺度宜人，内部充满了大小不一的广场与游戏场。近年落成的Plaça Lesseps广场原本是嘈杂的交通要塞与杂乱的环境，因此而生的居住与社会问题让人望而却步。新的广场将快速道路隐藏到地下，将地面变成可让人停留的空间。广场边的Jaume Fuster图书馆由建筑师Josep Llinás i Carmona设计。内部空间达5000m²，由两个量体将背后的山丘与Gracia街区连结为一体，面对广场的量体用屋顶与挑空让室内与室外产生互动。这个图书馆的出现为衰败的街区带来活力，吸引都市中产阶级与学生重新回到Lesseps广场。Gracia街区的居民活动也因而延伸到这里。

Cerdà的设计将公共空间视为第一优先，宽敞的街道成为一条条线性的生活空间。Passeig de St. Joan为近年完工的大道，由当地的建筑师Lola Domènech 设计。这条连接到凯旋门的大道宽达50m，以人行道与自行车道为主，两旁的植栽茂密，游戏场、小酒馆、街道家具等等的设计让这个大道变成居民可以真正生活的空间。

加泰隆尼亚广场为交通枢纽，大量的观光客让这里也成为重要商业中心。相连的La Rambla 大道从这里将旧城区一分为二，西为Raval区，东为Barrio Gotico。这里的空气里总是有着一股陈旧潮湿的味道，但在许多转角处，新设计的广场与公共设施总是带给你惊喜。Raval区中央的巴塞罗那当代美术馆与文化中心最具代表性。前者出自白色派教父Richard Meier之手，1995年落成启用。Meier的招牌白色量体与白色空间在地中海的阳光下映出迷人的光影。后者则为HelioPiñón 和 Albert Viaplana的作品，文化中心的概念则是用大面积的玻璃映照出周遭的历史建筑物，这个设计将旧的市议会建筑更新为文化中心，并定期举办各种文化活动与音乐会。两个公共设施极为完整地与周围的公共空间设计结合在一起。

广场上来往穿梭的是旅人、居民、学生、玩滑板的、拉小提琴的，如此刺激了当地的商业，带来更高的吸引力，尤其刺激了餐饮服务业的成长，也吸引许多艺术家来此定居。在外围小区，过去因建筑物残破与治安问题而使房价低迷，在整个公共空间重整之后，建设的资金因有市场价值而成长，房地产的价值也随之攀升。即使城区仍旧存在着复杂的社会问题，但在公共空间重整之后又恢复了生命力。

沿着看似无尽头的水岸走，沿线配置重要的空间如旧港、旧街区Barceloneta、奥运村，整个水岸的精华区最后以赫尔佐格·德·梅隆设计的世界文化论坛Forum Universal de la Cultura作为终点。巴塞罗那的水岸空间重新塑造了新的多功能城中心，为城市创造了新的形象。

可以说巴塞罗那的公共空间就是整个都市。把公共空间对城市的影响放在第一位的设计思想，使这座城市易于阅读，具有弹性，不同的街区因地景元素如丘陵与海洋，显现不同的性格。公共设施与公共空间，尤其是其中的公共文化空间，一方面使这里成为真正适宜居民生活的场所，另一方面则将城市整合为一体的有效工具。巴塞罗那的都市形象与居民的认同感建立在不可或缺的公共空间上。这个城市用规划传统与企图心创造出“巴塞罗那模式”，至今仍是其他城市竞争者学习的对象。

—

张道铭 & Ilaria Mancini　B-OND 建筑与都市设计工作室合伙人

Of all cities in the world, Barcelona has always been standing out with its fine buildings and extraodinary urban public space. The city, even now in the economic downturn, yet remains to be a mecca for global achitecture enthusiastists. The development of public works restructured the city`s space and brought changes to the lifestyle of citizens. The exquisitely handled public space also contributed to the leading position of Barcelona in the Europe in this field.

The planning of public space, involving both quality and quantity perspectives, was based on the framework of Eixample, which was put forward by IldefonsCerdà back in 1855, in response to the epidemic public higene problems due to the sudden increase of urban populations. The Eixample is a grid pattern, characterized by octagonal blocks with common inner yard and crossed by very wide streets with narrowest 20 meters. Despite that Cerdà`s design was somehow compromised when implemented, with the small squares being completely filled with additions by residents, the planning still gave considerations to public space and sunlight needs, ensuring fine living environment for citizens in the rapid urban expansion.

As the public space is considered to be part of infrastructure, the public works shall not be done merely for aesthetic reasons, but with the aim to renew abandoned areas in the city and to deal with complex social problems. The public interest was taken on board since the first general election held in 1979. And three leftist mayors in succession working on supportive policies, altogether, makes Barcelona as it is today.

The creation of public space in Barcelona roughly falls into three phases (Orio Clos, 2011). Phase I started in 1980s, when the municipal government tried to revive those areas in decay that were nearly forgotten by means of remaking public space and through small-scale public facilities –typically, the renewal of Ciutat Vella. The experimental policy was followed by initial success. Phase II refered to the preparation period for the Olympic Games, when most public tvansit and major public facilities for the metropolitan area were put in place and readily connected. Phase III started at the post-Olympics period and ended in 2010, when the city became more ambitious with the public works reaching out for northern and southern outskirts of Barcelona and large infrastructure having been built, such as the international airport and harbors.

The construction areas for Barcelona Olympics were integrated into the urban planning framework dating bake to the 1980s. There were four such areas. One for major sports venues, located at the southwestern Montjuic massif that was expected to become a fully functional urban garden; one was waterfront Olympic Village, that has helped citizens regain access to the sea by renewing waterfront space and ports; one was seated at the northwestern Vall d' Herbron with the aim to redevelop the suburban area in disorderly urbanization; and the last one was situated at the city proper –Diagonal, to rebuild the already existing public and sports facilities. Barcelona city government used the new public facilities to create the New Centrality (Juli Esteban, 2004) and to gain renewed vitality for the community. The construction for the Games meant not to satisfy international sports event only, but also provided solutions for then urban problems, such as decline, abandonment and urban spread.

The main venue for the Games was built in 1929 during the Barcelona Expo, but it didn't return in the spotlight after long time being forgotten and abandoned until Barcelona entered the bidding to be the host city of Olympic Games. Finaly, an international planning team, including Correa, Milà, Margarit, Buixadé and Gregoretti, won the bidding to renew the building. The neoclassic facade was wholy preserved, while the main structure and capacity were both redesigned to create an Olympic stadium that seats 56,000 people. And a new landmark was

本页：奥运村地标建筑群，一系列的公共设施与私人开发案，包括Frank Gehry设计的金鱼，让水岸空间成为城市的新中心。
对页，顺时针方向：Jaume Fuster 图书馆。Montjuic山丘的奥运跳水池，简洁的设计与丰富的光影变化，在通道即可远眺整个市区。水岸空间的公共艺术，邻近海岸街区Barceloneta。

This page: The landmark complex of the Vila Olimpica, a series of public facilities and private development projects, including the Goldfish designed by Frank Cehry, make the waterfront space a new centrality of the city (photo: Zhang Daoming).
Opposite page, clockwise: Jaume Fuster Library (photo: Zhang Daoming). The Olympic diving pool at Montjuic features a succinct design and abundant changes of lights and shadows. The aerial view of the city is available from the passageway (photo: Zhang Daoming). The public art at the waterfront space is near the coastal district Barceloneta (photo: Ilaria Mancini) .

designed by Calatrava in the square right before the main venue. Also worthy of note is the Olympic diving pool. With the magic of the architect Antoni De Moragas, the original swimming pool was renovated for Olympic purposes, and the design that takes the advantage of surrounding landscapes provides both audiences and athletes with the aerial view of the city of Barcelona. The entire public space of Montjiuc was replanned –the ramshackle buildings gained new birth, the accessibility to public transport increased (e.g. by way of bus and cable car the mountain top became reachable), and the urban development on the massif was contained. Thus green land was preserved in good amount, and the massif, in the real sense, was now a public garden that was accesssible for citizens.

Along Baixada and from north to south, you can find yourself in the continuous and integrated public space and in the interactions of public space.

The first thing is the square at the intersection in Gracia. The streets in Gracia are appropriate in both length and width, dotted with squares and entertainment parks of various sizes. The newly built Plaça Lesseps is located where used to witness bad traffic and mess and where people felt daunted to move in due to living and social problems. The square has the express way hidden underground and leaves ample space for people to linger in. Jaume Fuster Library nexr to the square was designed by the architect Josep Llinás i Carmona with the building floor totaled 5,000 square meters. Two buildings connect the massif and streets together, with the one facing the square using the roof and risen space to create interactions between the indoor and the outdoor. The advent of the library brings vitality to the declining community and attracts urban middle class and students back to Lesseps Square. And the community activities are also extended here.

The design of Cerdà always prioritized the public space, with the wide avenues and streets becoming linear living space. Passeig de St. Joan is a newly finished thoroughfare, designed by the local architect Lola Domènech. It ends in Arco de Trionfo and is 50m in width. Mainly designed for pedestrians and bicycles, it is decorated with luxuriant vegetation at both sides. The entertainment fields, taverns and street furniture, altogether, make the thoroughfare a truly livable space.

Plaça Catalunya is both a traffic center and an important commercial center for tourists. Standing at the plaza and along La Rambla Boulevard, the west side is Raval and the east Barrio Gotico. The air here is fusty, misty and moisty, but the new square and public facilities always bring surprises in the corners, typical examples being MACBA and CCCB in the center of Raval. MACBA, designed by Richard Meier (the Whites), was opened in 1995. Meier`s signature white building and white space cast beautiful light and shadows in the Mediterranean sunshine. CCCB was co-designed by HelioPiñón and Albert Viaplana. The idea was to have surrounding historical buildings reflected in the facade glass in use. The cultural center was built by reforming the old hall of city council, where cultural activities and concerts happen regularly now. These two public facilities have been well integrated into the public space around.

On the plaza travel the tourists, residents, students, skateboarders and violinist, which provides a driving force for commercial activities and the catering industry and attracts many artists to settle here. On the periphery, the housing price used to slump due to the conditions of buildings and security problems. After the renewal of the public space, the construction fund increases with the market trend and the housing price also rises. Although there are still complex social problems needing to be solved, vitality was regained in the old districts.

The seemingly endless waterfront is lined with important space, such as Port Vell, Barceloneta, the Olympic Village (Ciutat Olimpica) and the highlights end with the World Cultural Forum (Forum Universal de la Cultura) designed by Herzog De Meuron. The waterfront space remakes a new functional city center for Barcelona, creating a new image for the city.

It is fair to consider the public space in Barcelona as the city as a whole. The design idea of prioritizing the impact of public space on Barcelona makes the city ever more accessible and resilient. Different communities show different personalities, depending on various landscape factors, such as hills and the sea. The public facilities and public space, particularly the public cultural space, on the one hand creates a genuinely livable space, and on the other hand, effectively integrates the city into one being. The city image of Barcelona and residents identifying with the city are based on the indispensable public space. Barcelona has produced a Barcelona Pattern by following the planning tradition with ambition, and remains to be the trend-setter even today.

—

Text: Zhang Daoming and Ilaria Mancini, B - OND architecture and urban design studio partners

Welco

曼彻斯特的市中心公共空间改造

Transformation of Public Space in the Downtown of Manchester

曼彻斯特所进行的多项重要的都市再开发案，努力联结既有的都市纹理，并赋予曼城一个新的面貌，使其成功地转型为一个包含文化产业、娱乐消费与休闲生活的中心。

Manchester conducts a series of important urban planning of redevelopment to link with existing urban texture, making Manchester city a new look, successfully transformed into a centre containing the functions of cultural industry, entertainment, and leisure life.

Text
蒋文德

在20世纪30年代的经济大萧条时期，曼彻斯特的纺织工业迅速衰败凋零，人口逐渐衰减。到了20世纪中叶，曼彻斯特的工业已失去了重要性，逐渐在1990年代转型为一个以商务、观光、休闲娱乐、媒体与文化产业为主的后工业城市。

在现代都市设计的影响下，上世纪90年代后期曼彻斯特所进行的多项重要的都市再开发案，企图对衰败的市中心重新投资，努力联结既有的都市纹理，并赋予曼城一个新的面貌。在再开发的过程中，吸引了大量促进城市缙绅化的相关投资，原来工业时代遗留下的运河、工业建筑与其他重要的公共建筑与历史建筑，形成了空间消费与缙绅化过程中的重要文化资产。而1996年爱尔兰共和军恐怖袭击对于市中心造成破坏后，该地区则是先经由审慎的总体规划竞图，再进行多项重建与修建计划。如今曼彻斯特成功地转型为一个包含文化产业、娱乐消费与休闲生活的中心。

此外，曼彻斯特市的城市空间结构，是以市中心的商业、娱乐与住宅混合区为主，并以放射状的地面电车系统形成主要的公共交通网络，联结周边较低密度的住宅区，以及大曼彻斯特区的其他城镇。便捷的电车交通网络，与符合人性尺度、适宜徒步活动的市中心，促使曼城的中心商业区不仅仅满足办公商务活动，它也是整个大曼城区城市居民休闲娱乐与文化活动的中心。与国际都会伦敦或巴黎相比，曼彻斯特的市中心更像是大曼城区市民生活为主的活动中心，因此曼彻斯特的市中心改造，似乎更围绕着城市居民的休闲生活。

在曼城的市中心，Piccadilly Gardens是其中最大、最主要的广场，它位于城市主要的旅馆区与零售商业区之间，距离Piccadilly 车站十分钟步行路程。在90年代末重新规划改建前，Piccadilly Gardens是一个低于路面的维多利亚式花园，由于缺乏与街道视觉上的直接联结，道路与电车轨道形成了街道活动与广场空间的阻隔，再加上当时经济的衰落，Piccadilly Gardens俨然形成了一个缺乏生气的遗弃空间，并成为犯罪的温床。而新的Piccadilly广场，由EDAW与安藤忠雄规划设计，试图在周边临近的中国城、运河街同志村、零售商业区、电车转运站，与新建的千禧特区之间，创造一个 24小时使用的公共开放空间。

首先，广场的北面利用人行徒步区与花台植栽重新连结上周边具有历史性的街廓空间，东南面则用一栋七层建筑阻隔繁忙交通的噪音干扰，并提供地面层的人行穿越动线与零售餐饮活动。广场的南面与西面则由安藤忠雄设计的弧线造型的凉亭咖啡与商店，界定了是广场的活动界面，并提供适当的穿透性与阻隔性，一方面连结凉亭另一侧面的电车车站，也同时减缓电车轨道与噪音对广场的影响。广场中心的椭圆形喷水池，是整个广场的视觉焦点，贴于地面水池的设计，提供了亲水游戏的休闲气氛。而喷水池上南北向的步道，则连接了广场北面街道与广场南面的电车车站。

因此，整个广场的空间设计，除了广场内的视觉焦点外，也包括了与临近街廓的空间联结，周遭人行动线与

活动的结合，以及广场界面的妥善设计。成功的空间规划设计，妥善地处理了交通节点的正面与负面影响，塑造了半围闭的、慵懒的开放空间，广场上各项活动错落安排，包括周末的临时市集与户外游乐场，以及街道活动与人行动线带来24小时全天候的自然监管功能，使Piccadilly Gardens成为曼城市中心最受欢迎的广场，一个城市居民的活动中心。

除了Piccadilly Gardens，千禧年前后的曼彻斯特，市中心的另一项重大公共空间的投资就是千禧特区的再开发案。千禧特区的开发，一方面是因1996年IRA爆炸案对于曼城市中心的破坏，另一方面更试图利用绿地、广场与行人徒步区整合现代的零售空间、文化景点与历史建筑，提供一个结合休闲、消费、历史、著名建筑与开放空间的综合开发案，以其强化曼彻斯特的城市风貌，并活化城市中心的商业发展与空间活动。

千禧特区主要包括两个不同特色的广场，首先是位于曼彻斯特玉米交易所(Corn Exchange，现为商场)、Arndale购物中心与Selfridges 百货公司三栋建筑所形成的一个三角形开放空间 ——交易广场(Exchange Square，现为The Triangle Shopping Centre)。另一个则是玉米交易所另一侧与曼彻斯特教堂和足球博物馆(National Football Museum，之前是Urbis Museum)所围闭的教堂广场交易广场是以一种非正式的空间形式，透过层叠的弧形坡道矮墙提供户外剧场形式的座椅，成为逛街民众暂时歇息的据点， 或是上班族午餐休息的场所。广场的底层以填满石板的人造溪流作为视觉焦点，历史建筑玉米交易所的立面作为“舞台背景”，与其西侧16世纪的木桁架建筑，提供了空间的历史遐想。在玉米交易所历史建筑立面上所悬挂的大屏幕，与其西侧古老建筑里的酒吧，吸引大量球迷群众聚集，是周末城市居民观赏球赛的好去处。

相较于交易广场，玉米交易所另一侧的教堂广场，则是以轻松悠闲的现代雕塑与绿地景观为主，提供足球博物馆与曼彻斯特教堂一个清幽宁静的‘后花园’，并有效联结了古老厚重的歌德教堂与现代具半穿透感玻璃建筑物。相较于交易广场的都市特质，教堂广场为这个城市最古老的区域提供了一个舒缓休闲的空间气氛。

除了市中心的重要广场与商业消费活动之外，曼彻斯特艺文生活与会展活动则集中在曼彻斯特市政中心周边的美术展览馆，中央图书馆，歌剧院与Bridgewater Hall音乐厅。

曼彻斯特市政厅是一座维多利亚时期所建的哥德复兴式建筑，市政厅前西面的阿伯特广场上矗立着维多利亚女王夫君阿伯特亲王的纪念碑。市政厅南侧的增建部分与中央图书馆是1930年代所新建的折中主义与古典主义建筑，并将市政中心周边较为严肃的商业办公氛围延伸并转换至南面以活动为主的曼彻斯特会展中心、曼彻斯特歌剧院与Bridgewater音乐厅。建于20世纪初的曼彻斯特歌剧院，为曼城提供了芭蕾、音乐剧与其他剧院类型的巡回表演。而Bridgewater音乐厅则建于90年代，是The Hallé交响乐团的驻所，也是BBC爱乐的主要表演场所。

曼彻斯特从一个工业革命兴起的工业城市，历经20世纪后的衰败，逐渐转型为一个以服务、文化产业为主的城市。市中心的公共空间的投资，也随着对于城市活化与城市再开发的概念，结合了城市的历史元素、居民的生活方式与后现代的消费文化，提供了更广阔且更多样的公共空间与文化活动。尽管城市缙绅化仍是一个可争议的议题，但曼彻斯特在上世纪90年代末期起展开的一连串都市公共空间的开发案，基本上在尊重城市历史脉络，与强化城市多元文化的条件下，提供了更好的都市公共空间质量，丰富的文艺活动与展场空间，以及一个更适合人居的都市生活环境。

—

蒋文德 伦敦政治经济学院城市设计师与社会科学硕士，原伦敦黑克尼自治市资深都市设计师

开篇：足球博物馆与曼彻斯特教堂（Manchester Cathedral）之间的教堂广场（cathedral Square），以现代雕塑与绿地景观为主，提供较轻松悠闲的开放空间。

跨页及对页：千禧年后重建的Piccadilly Gardens是曼城市中心最大、最受欢迎的公共开放空间，周末的市集与暂时性的游乐设施，以及24小时使用的规划设计，使Piccadilly Gardens成为曼城市民的生活活动中心。

本页下：位于Salford City的英国媒体城，距离曼城市中心只有十五分钟电车车程，为近年曼彻斯特文化产业发展中最重要的开发项目，并提供曼城居民一个观赏现场电视制播或是周末休闲的去处。

Opening page: Cathedral Square situated in between Football Museum and Manchester Cathedral, is a relaxed and leisurely open space with modern sculpture and green spaces.

Spread page and Opposite page: Piccadilly Gardens rebuilt after millennium is the largest and most popular public open space of the city, having the weekend bazaar and temporary amusement facilities, as well as the planning and design for 24-hour make Piccadilly Gardens become a living and activity centre for residents (photo: en.wikipedia.org).

This page: MediaCityUK located in Salford City, 15 mins by tram to downtown of Manchester, is one of the most important projects for Manchester culture industry development recent years, and a wonderful place for residents of Manchester to watch live TV production and spend leisure time on weekend.

本页：曼彻斯特市政厅前的阿伯特广场，矗立着维多利亚女王夫君阿伯特亲王的纪念碑. 与Piccadilly Gardens比较，阿伯特广场周遭为较严肃的行政与商业办公活动。

对页： 玉米交易所旁的Old Wellington Inn and Sinclair's Oyster Bar，是曼城最受欢迎的酒吧之一。其历史可追溯至16世纪都铎时期。在20世纪90 年代IRA爆炸案之后，新建的Shambles Square 重建了曼城这间最古老的木构造桁架建筑。

This page: Albert Square in front of Manchester City Hall stands a monument to Prince Albert, the husband of Queen Victoria. Compared with Piccadilly Gardens, the surrounding of Albert Square is a more serious administrative and office area（photo:upload.wikimedia.org）.

Opposite page: Old Wellington Inn and Sinclair's Oyster Bar next to Corn Exchange is one of the most popular bars in Manchester, Its history can be traced back to the 16th century of Tudor times. After IRA bomb in 90s, a new Shambles Square was rebuilt to present the oldest wood structure building.（photo: toms-travels.net）

During the Great Depression in the 1930s, Manchester' s textile industry declined rapidly, and the population decreased gradually. By the mid-20th century, Manchester's industry lost its significance, and it was transformed into a post-industrial city focusing on business, tourism, entertainment, media and culture industry in the 1990s gradually.

Under the influence of modern urban design, various important urban redevelopment projects proceeded in Manchester in the late 1990s tried to reinvest and reconstruct Manchester regarding the downtown decay and the damages caused by IRA's terrorist attack in 1996, strived to bind existing urban fabrics, and gave Manchester a new look. The canals, industrial buildings and other important public buildings and historical buildings left over in the original industrial age become important cultural heritages in the process of space consumption and gentrification. Through prudent competition on overall planning, and after conducting many reconstruction and construction plans, nowadays, Manchester has transformed into a center covering culture industry, entertainment, consumption and leisure successfully.

In addition, Manchester's urban spatial structure relies on the mixing areas of business, entertainment and residence in the downtown, and forms the main public transportation network with radial ground tramcar system to connect surrounding residential areas with low–density and other cities and towns in the Greater Manchester. The convenient tramcar transportation network and the downtown conforming to human scale and being suitable for pedestrianism activities enable Manchester's central business district to meet the demands of office and business activities, and become the center of entertainment and cultural activities for residents in the whole Greater Manchester as well. Comparing to cosmopolitan city London or Paris, the downtown of Manchester is more like an activity center focusing on the citizen life of the Greater Manchester. Therefore, it seems that the transformation of Manchester's downtown centers on the causal living of urban residents.

In the downtown of Manchester, Piccadilly Gardens is the biggest and most important square. Situated between the main hotel area and retail area of the city, it is 10 minutes' walk from Piccadilly station. Before the re-planning and reconstruction in the late 1990s, Piccadilly Gardens was a Victorian garden below the road surface. Due to lacking of direct connection with streets visually, roads and tram rails separate street activities and square space. In addition, the decline of economy at that time made Piccadilly Gardens a lifeless abandoned space, a hotbed of crime. While the new Piccadilly Gardens, planned and designed by EDAW and Tadao Ando, tried to create a 24-hour public open space among the adjacent Chinatown, Canal Street, retail area, tramcar station and the newly-built Millennium Quarter.

Firstly, the northern side of the square re-connects the peripheral historic block space by walking area and parterres, while in the southeast, a seven-story building blocks the noise disturbance from busy traffics and provides the ground-level pedestrian route and retail and catering activities. In the southern side and western side of the square are curve-shaped coffee gloriette and stores designed by Tadao Ando, defining the activity boundary of the square and providing proper penetrability and barrier. It links to the tramcar station in the other side of gloriette and reduces the effects of tram rail and noise. The oval fountain pool in the center of the square is the visual focus of the whole square, and the design of pool on the ground provides the casual atmosphere of water loving and playing. And the north-south footpath upon the fountain pool connects to the streets in the northern side of the square and the tramcar station in the southern side of the square.

Therefore, the space design of the whole square, in addition to the visual focus in the square, also includes the space connection with nearby blocks, the combination of pedestrian route around and activities and the proper design of square boundary. It is a successful space planning and design which handles the positive and negative effects of transport nodes properly and creates a half-closed and lazy open space. Various activities are arranged scatteringly in the square, including weekend temporary market and outdoor playground. Street activities and 24-hour natural monitoring function brought by pedestrian route as well make Piccadilly Gardens a most popular square in the downtown of Manchester, an activity center for urban residents.

Besides Piccadilly Gardens, the other important investment in public space in Manchester before and after 2000 is the redevelopment project of Millennium Quarter. The purposes of the development of Millennium Quarter are to respond to the damages of IRA explosion on the downtown of Manchester, and make use of green space, square and pedestrian area to integrate modern retail space, cultural sights and historical buildings. A comprehensive development project combining leisure, consumption, history, famous building and open space is provided to reinforce the city-scape of Manchester and activate commercial development and space activities in the city center.

Millennium Quarter includes two different squares, namely, the Exchange Square (being the Triangle Shopping Centre now), a triangular open space formed by Manchester Corn Exchange (being a store now), Arndale Center and Selfridges, and the Cathedral Gardens enclosed by the other side of Corn Exchange, Manchester Cathedral and National Football Museum (Urbis Museum formerly).

Exchange Square, in a form of informal space form, provides outdoor theatre seats with arc-shaped ramps and parapets, becoming a place for the masses who are shopping to have a rest temporarily or for office workers to have lunch or have a rest. The ground floor of the square takes the artificial stream filled with slabs as the visual focus and the façade of the historical building Corn Exchange as the "stage background" . Together with them, the wood truss building from the 16th century in the western side of Corn Exchange provides historical imagination of space. The large screen hanged in the façade of Corn Exchange and the pub in the western old building attract a large number of fans and masses, being an ideal place for urban residents to watch the matches on the weekends.

Comparing to Exchange Square, Cathedral Gardens in the other side of Corn Exchange, focusing on leisure modern sculptures and green space landscapes, provides National Football Museum and Manchester Cathedral a quiet and beautiful "back garden", and combines the old and massive Gothic church and modern trans-reflective glass building effectively. Comparing with the urban characteristics of Exchange Square, Cathedral Gardens provides a leisure and casual space atmosphere to the oldest area in this city.

Besides the important squares and commercial activities and consumptions in the downtown, Manchester's activities regarding art, culture, living and exhibition are concentrated in Manchester Art Galley which is close to Manchester Town Hall, Manchester Central Library, Manchester International Convention Centre and Manchester Central Convention Complex, Manchester Opera House and Bridgewater Hall.

Manchester Town Hall is a Gothic Revival architecture built in Victorian era. In Albert Square which is in the front and western side of Town Hall, stands the monument of Prince Albert, the husband of Queen Victoria. The annex and Manchester Central Library in the southern side are eclectic and classical architectures built in the 1930s, extending and converting the serious business and office atmosphere surrounding Town Hall to Manchester International Convention Centre and Manchester Central Convention Complex, Manchester Opera House and Bridgewater Hall which focus on activities in the northern side. The Manchester Opera House built in the early 20th century, provides tours of ballet, musical and other theatre categories to Manchester. While the Manchester Opera House, built in the 1990s, is the place where The Hallé symphony orchestra stays and the main arena for BBC Philharmonics.

Manchester, after the decline in the late 20th century, was transformed into a city focusing on service and culture industry from an industrial city emerging in the Industrial Revolution. The investment in downtown public space, along with the concepts of urban renewal and urban redevelopment, combines the historical elements of the city, the lifestyle of residents and post-modern consumer culture and provides more and multiple public space and cultural activities. Even though urban gentrification is still a controversial issue, a series of urban public space development projects proceeded from the late 1990s, under the conditions of respecting urban historical context and strengthening diverse urban culture, provide better urban public space, rich cultural activities and exhibition spaces and a more hospitable urban living environment.

—

Text: Jiang Wende, Master of Graduate Institute of Building and Planning, NTU, urban designer and Master of Social Science of the LSE, former senior urban designer in LBH

布法罗的公共空间规划

Buffalo's Public Space Planning

布法罗的中心区复兴计划将重心放在了改建或修复已有的建筑遗产上，使其焕发了新的活力。并且在2006年以来，以实施“水岸走廊计划”强调统一协调和连接沿岸景观、公园和其他休闲设施。

The revival plan of buffalo center focuses on rebuilding or repairing existing architectural heritage, making it a new vitality. Since 2006, government has implemented the "plan of corridor in the water bank", which emphasizes the unified features and coordination of landscapes, parks and other recreational facilities along the coast.

Text
卢端芳

开篇：海军军事公园经过多年发展，已成为拥有多艘军舰以及军机、坦克和潜艇的露天博物馆，是水岸老少咸宜的观光点。
本页：拉斐特广场的雕像。
对页：经成功改建，谢亚表演艺术中心已重新成为纽约州最成功的剧场之一，是纽约百老汇音乐剧巡演的重要一站。
Opening page: Naval and Military Park（Photo: farm9.staticflickr.com）. After several years' development, it has become an open-air museum owning many warships, military plans, tanks and submarines, and an attraction in the water bank for all age.
This page: The statue in lafayette Square.
Opposite page: Shea's Performing Arts Center. After successful reconstruction, Shea's Performing Arts Center has become one of the most successful theatres in New York State again and an important site of New York Broadway musical tours（Photo: farm9.staticflickr.com）.

临近美加边界尼亚加拉瀑布的布法罗现有人口113.5万，它既是纽约州继纽约之后第二大城市，也是五大湖区继芝加哥之后第二大城市，因此，“皇后之城”成为该市最广为人知的绰号。三百年前的布法罗曾是一个野牛聚集、印地安人自由奔走狩猎之地。18世纪中叶，法国皮货商率先在此建立皮毛贸易站，但在1763年法印战争后，法国人失势，来此定居的荷兰人逐渐增多，在布法罗河与伊利湖的交汇点附近建立起了名为“新阿姆斯特丹”的聚居点。1825年伊利运河开通后，布法罗迅速发展，成为美国五大湖区工业基地的重要组成部分。籍由尼亚加拉瀑布所提供的巨大能量，布法罗在20世纪初大兴发电业，成为美国第一个大规模提供电路灯的都市，因而亦有“光城”之称。20世纪下半叶，随着圣劳伦斯水道的开通，布法罗的交通枢纽地位不再，钢铁、面粉加工等工业也迁至它地，经济自此一蹶不振。布法罗与五大湖其他老工业基地如底特律、克里夫兰一道，被视为衰落的锈带，及至1990年，其人口已锐减至1900年的水平。进入21世纪，布法罗致力于发展医疗、教育和零售业，其生物讯息及人类基因研究中心罗致了一批世界顶尖人才，零售业也成功吸引大批加拿大消费者来此购物，经济得以逆流稳步增长。

布法罗市中心的起伏跌荡可谓是整个城市兴衰的缩影。1804年，测量师约瑟夫·埃利科特来到当时尚是一个小村庄的布法罗，为其作了规划，整个市中心的空间架构自此基本成型。其时正值美国首都华盛顿营建之初，埃里考特的方案受之影响颇深，城市功能让位于空间形式，强调轴线关系及良好的视觉空间结构。在整个城市的骨架方面，他首先确立的是两个主要公共广场及以其为中心呈轮辐状向外发散的主要道路。临近伊利湖的尼亚加拉广场是八条主街的交汇点，也是整个城区的中心。在城市发展的起步阶段，它的四周曾是最具吸引力的居住地点。到了20世纪30年代，随着市政厅大厦、州政府办公大楼和联邦法庭的陆续建成，尼亚加拉广场周边区域成为布法罗的行政办公中心。高达32层的市政厅，不但规模宏大，还是全美最高的市府大楼之一，并且以繁复华丽的装饰派艺术细部，淋漓尽致地呈现了布法罗在其鼎盛时期的辉煌。站在市政厅巨大的彩色装饰圆拱顶下抬望，犹可想见当年该市踌躇满志、“天下舍我其谁”之气势。市政厅大厦正立面气势恢弘的对称设计，强化了市中心的东西轴线，并成为尼亚加拉广场标志性的景观。广场四边其他建筑则以变化多端的形体，对广场起了很好的围合作用，却并不喧宾夺主。广场的中央高高耸立着纪念1901年来布法罗参加泛美博览会时遇刺身亡的美国前总统威廉·麦金利的方尖碑，成为整个广场以及周边主街的视觉焦点。

由市政厅大厦与方尖碑所确立的东西向视廊具有极强的导向感，很好地强调了尼亚加拉广场与周围环境在景观与形体上的联接。沿着这个视廊往东走三个街区，很快就来到布法罗市中心的另一主要公共空间——拉斐特广场。1825年法国将军拉斐特在他历史性的美国之旅中参观这个广场并发表讲话，拉斐特广场自此得名。在很长一段时间里，这里曾是树木繁茂的城市公园。随着交通需求的增长，公园几经重建，面积逐渐缩小。与气氛肃穆的尼亚加拉广场相比，尺度小巧的拉斐特广场更富有亲和力。每年夏季从5月到9月，都会有一系列免费音乐会在此举行，使得拉斐特广场成为市民最喜爱的消夏场所之一。拉斐特广场的周边是布法罗的金融中心，云集了十几家银行的办公楼，包括由著名的日裔美国建筑师雅马萨奇设计的M&T银行。图书馆、剧场、教堂、画廊等场馆也散布其中，整个区域弥漫着浓郁的文化气息。

营建于城市发展鼎盛时期的布法罗市中心布局大气，建筑精致，可惜不少昔日荣华而今只遗一副躯壳：市区大半办公楼空空荡荡，益发显得大而不当。因此，与其他城市热衷于兴修昂贵的奇景建筑不同，布法罗的市区复兴计划将重心放在了改建或修复已有的建筑遗产上。以测量师埃利科特命名、面积41,500m^{2}的埃利科特广场大厦在1896年建成之初曾是世界上最大的办公楼，其大堂经过改建，现已成为提供民众聚会、婚宴庆祝等多种功能的公共空间。谢亚表演艺术中心的起死回生则是较早期的一个成功案例。建于1926年的谢亚表演艺术中心，以17–18世纪的欧洲宫廷剧院为原型设计而成，细部糅合了西班牙和法国的巴洛克和洛可可风格，不少元素今天看来还很到位。20世纪60到70年代，随着布法罗的日益萧条，剧场也面临破产。由于业主持续拖欠高昂税费，法庭将剧场判给了市府和一直在支持剧场修复的社会公益志愿者团体。经成功改建，谢亚表演艺术中心已重新成为纽约州最成功的剧场之一，是纽约百老汇音乐剧巡演的重要一站。丰富的文化休闲生活是吸引优质人才的必要条件，而对建筑遗产的再利用不但能以较小的成本提供高质量的公共空间，并且能使城市的历史脉络得以延续，拥有更为深远的内涵，值得我们更多的城镇仿效。

布法罗城市发展的一个主要缺陷是：虽然市中心紧临河湖，却被繁忙的高速路所阻隔，使得面积77公顷的水岸虽近在咫尺，却未能成为都市公共空间的有机延续。并且，大部分滨水区过去为工业基地，缺乏相应的休闲设施。近年来，随着布法罗城市转型的深入，滨水区的再发展成为规划重点。在2006年制定的“皇后之城2030年总规”中，“水岸走廊计划”强调连接和统一协调沿岸景观、公园和其他休闲设施，使居民得以更好地享受布法罗得天独厚的自然环境。近年来，水岸复兴零敲碎打地进行着。始建于1977年海军军事公园经过多年发展，已成为拥有多艘军舰以及军机、坦克和潜艇的露天博物馆，是水岸老少咸宜的观光点。由于城区与滨水区的连接尚不顺畅，目前，非营利组织正游说相关部门将5号公路线改造成为贯通城西与水岸的林荫大道。

都市空间的营造并非一蹴而就，需要在岁月的流光里依照城市自身的节奏缓慢冲磨。如老子所云，“孰能浊以止，静之徐清？孰能安以久，动之徐生？”—— 新的城市文化的生成当如大自然静息中从容的微动，在混沌中逐渐清朗。相信布法罗在尽得河湖之利后，能在21世纪涅磐重生，为其他衰败都市的复兴提供一个合适的样本。

—

卢端芳，澳大利亚悉尼大学建筑设计与城市规划学院副院长、教授，澳大利亚国家研究理事会“未来院士奖”学者

SHEA'S
BUFFALO
THANK YOU SHEA'S SPONSORS
Wegmans
SHEA'S
PERFORMING ARTS CENTER
ACQUEST
SHEA'S TICKET OFFICE

Located near Niagara Falls which straddle the international border between Canada and the United States, Buffalo, with a population of 1,135,000, is the second biggest city not only in the state of New York, after New York City, but also in the Great Lakes Region, after Chicago. Therefore, "Queen City" became the best known nickname of this city. About 300 years before, Buffalo was a place where buffalos gathered and Indians ran and hunted freely. In the mid-18th century, French fur traders were the first to establish fur trading posts here. However, after the French and Indian War in 1763, French lost their powers and influences, while Dutch who came here for settlement increased and built a settlement named "New Amsterdam" near the intersection of Buffalo River and Lake Erie. After the opening of the Erie Canal in 1825, Buffalo grew quickly and became an important part of Great Lakes Industrial Base. At the dawn of the 20th century, Buffalo developed power generating industry energetically via the tremendous energy provided by Niagara Falls, and became the first city in America deploying electric street lights on a large scale. Consequently, the city got the nickname "City of Light". In the second half of the 20th century, with the opening of the St. Lawrence Seaway, Buffalo lost its status as a transportation hub and moved steel, milling and other industries to other places, making its economy begin to deteriorate. Together with other old industrial bases in the Great Lakes Region, for example, Detroit and Cleveland, Buffalo was seen as the declining Rust Belt. By 1990 the city had fallen back below its 1900 population levels. In the 21st century, Buffalo focuses on developing medical treatment, education and retail industry. Its biological information and human genome research center recruits a group of world top talents, and retail industry attracts a great number of Canadian consumers to do the shopping here, making its economy grow steadily.

It may be said that the ups and downs in the downtown of Buffalo are the epitome of the rises and declines of the whole city. In 1804 when the construction of Washington D.C, the capital of the United States, begins, the surveyor Joseph Ellicott came to Buffalo which is still a small village at that time and made a plan for it. Since then, the spatial structure of the whole downtown formed basically. Ellicott's scheme where urban function gives way to space form and axis relationship and good visual space structure are stressed was deeply influenced by Washington. In the framework of the whole city, he firstly determined two main public squares and main streets diverging outward in spokes. The Niagara Square near Lake Erie is the intersection of eight streets as well as the center of the whole urban area. In the beginning stage of urban development, places all around it were the most attractive residential places. In the 1930s, with the establishments of City Hall, office building for state government and federal court in succession, the perimeter zone of Niagara Square became the administration and office center of Buffalo. The 32-story City Hall is one of the highest city halls in the United States for its grand scale, and presents Buffalo's glories in its heyday vividly with complicated and magnificent details of art deco. Seen from the huge color decorative dome of the City Hall, we can still imagine the complacent momentum of "if I can't do it, who can". The grand symmetric design in the front elevation of City Hall reinforces the east-west axis of the downtown, and becomes a landmark of Niagara Square. Other buildings around the square, in variegated forms, enclose the square well, being not distracting. In the center of the square stands the obelisk commemorating William Mckinley, former US president who was assassinated in 1901 when he came to Buffalo to attend the Pan-American Exposition. It becomes the visual focus of the whole square and peripheral main streets.

The east-west visual corridor established by City Hall and obelisk has a strong orientation, stressing the connection of Niagara Square and surroundings in landscape and shape well. Walk 3 blocks to the east from this corridor to go to Lafayette Square, the other main public space in the downtown of Buffalo. In 1825, the French General Lafayette visited this square and made a speech in his historic American journey, and Lafayette Square became well-known since then. In a long time, here was a wooded city park. With the increase of transportation demand, the park downsized gradually after reconstructions. Comparing to the solemn Niagara Square, the smaller Lafayette Square is friendlier. From May to September in summer every year, there will be a serious of free concerts, making Niagara Square one of the favorite places for residents to spend the summer at leisure. The periphery of Niagara Square, being the financial center of Buffalo, gathers the office buildings of more than 10 banks, including M&T bank designed by Yamasaki, a famous American architect of Japanese origin. Library, theatre, church, gallery, and other venues also scatter here. The whole area is filled with full-bodied cultural atmosphere.

The Buffalo downtown built in the heyday of urban development was elegant and delicate. However, many splendors in former days become a shelf: most of the buildings in the downtown are empty, which looks improperly huge. Therefore, being different from other cities which are wild about building costly wonder buildings, the revival plan of Buffalo's downtown focuses on reconstructing or restoring existing architectural heritages. The Ellicott Square Building, named after surveyor Ellicott and with an area of 41,500 square meters, was the biggest office building in the world in 1896 when it was built. Its lobby has become a public space providing assembly, wedding banquet, celebration and other functions to the masses after reconstructions. The revival of Shea's Performing Arts Center was a successful case in the early stage. Built in 1926, Shea's Performing Arts Center was designed with the prototype of European court theatre in the 17-18 century. Mixing the Spanish and French Baroque and Rococo styles in the details, it can accommodate more than 3,000 audiences. The interior design was completed by the famous designer Louis Comfort Tiffany, and today it seems that many elements are still sufficient. From the 1960s to 1970s, with Buffalo's increasing depression, theatres faced bankruptcy. Due to the fact that the owner defaulted on high taxes and dues continuously, the court awarded the theatre to the Government and non-profit community groups who support the restoration of theatre all the time. After successful reconstruction, Shea's Performing Arts Center has become one of the most successful theatres in New York State again and an important site of New York Broadway musical tours. Rich cultural and leisure life is the requirement of attracting quality talents, while the reusing of architectural heritage can not only provide high-quality public space at lower cost, but also enable the city to continue its historical context and possess more profound contents, deserving the imitation of more cities.

A major defect in Buffalo's development is that even though the downtown is next to rivers and lakes, it was separated by busy highways, making the 77-hectare water bank nearby fail to become the organic continuation of urban public space. In addition, most of the waterfront was industrial bases in the past, lacking corresponding recreation facilities. In recent years, with the deepening of Buffalo's urban transformation, the redevelopment of waterfront become a planning emphasis. In the "overall planning of Queen City in 2030" formulated in 2006, the "plan of corridor in the water bank" emphasized the connection and coordination of coastal landscapes, parks and other recreation facilities, enabling residents to enjoy Buffalo's advantaged natural environment better. In recent years, the revival of water bank is conducted in dribs and drabs. Built in 1977, Naval & Military Park, after several years' development, has become an open-air museum owning many warships, military plans, tanks and submarines, and an attraction in the water bank for all age. Owing to the fact that the connection between urban area and waterfront is not so smooth, at present, Non-Profit Organization is lobbying relevant departments to transform the No.5 road line into a boulevard running through the west of the city and waterfront.

The construction of urban space can not be completed overnight. It needs to be shaped in years according to the city's own steps. As Laozi mentioned, "Who can make the muddy water still and become gradually clearer? Who can act in peace and make it vital?" – The creation of new urban culture should become gradually clearer like the deliberate final motion in the tranquillization of nature. It is believed that after Buffalo gains the benefits of rivers and lakes, it can get its rebirth in the 21st century, and provide an appropriate example for the revival of other declined cities.

—

Text: Lu Duanfang, Vice-president and Professor of Building Design and Urban Planning School of University of Sydney, scholar of "Future Academician Award" of Australian Research Council.

本页：从左到右：以测量师埃利科特名字命名的、面积41500平方米的埃利科特广场大厦在1896年建成之初曾是世界上最大的办公楼，其大堂经过改建，现已成为提供民众聚会、婚宴庆祝等多种功能的公共空间。布法罗市政厅圆拱顶和壁画。

对页：拉斐特广场正立面。

This page, from left to right: The Ellicott Square Building, named after surveyor Ellicott and with an area of 41,500 square meters, was the biggest office building in the world in 1896 when it was built. Its lobby has become a public space providing assembly, wedding banquet, celebration and other functions to the masses after reconstructions;Vaults and murals of Buffalo's City Hall.

Opposite page: Facade of Lafayette Square.

周继春，昆山城市建设投资发展有限公司董事长

Zhou Jichun, chairman of the board of Kunshan City ConstructionInvestment & Development Co., Ltd.

Interview with Zhou Jichun

Spreading Art and Reproducing Tradition

传播艺术和繁衍传统的土壤·周继春

新印象：文化艺术中心的项目筹建阶段各方有什么不同看法？

周继春：这个项目原先计划在已拆迁的昆山大剧院原址上翻盖一个剧院，然而在邀请了上海大剧院的负责人和华东院的相关专家进行论证时，专家的意见是：这里比不了苏州、上海，没必要盖大剧院，盖个文化中心就可以。但是昆山市的领导不这么看，认为昆山是昆曲的发源地，是有深厚的精神文化底蕴的江南灵秀之地，市民有欣赏高雅艺术的基础。因此最后拍板决定：文化艺术中心既要有影院和会议功能，更要包含昆曲表演场馆和大剧院。

新印象：昆山城投公司在什么阶段介入的？您当时对于文化艺术中心的筹建有什么思考和计划？文化艺术中心的自我定位和设计团队是怎样形成的？

周继春：昆山城投公司接手的时候，艺术中心已经完成了项目设计竞标。当时我个人是有顾虑的，虽然昆山的财政收入可以支撑这个项目，但项目会在一个什么样的轨道上运行，后期将呈现出一种什么样的面貌？必须进行有效的控制，否则难以保证建造一个高质量的作品。为此我们采用了一系列措施，首先是让我们的公司团队对有代表性的大城市剧院目前的现状进行充分的考察，带着问题和目的去实地了解上海、苏州、中山、杭州，武汉、重庆、深圳、北京等城市的剧院设计和运营情况。

根据考察的结果，我们明确了应该以一种什么状态来面对，哪些经验值得我们推崇，哪些问题我们应该避免，同时该组建什么样的各类专业团队来实现设计、管理设计并最终良性运营。最终我们希望它是在可控的良性轨道上有序的运转。在此我个人觉得中山市艺术中心是文化艺术界人士牵头建起来的，但它的运营模式很好，值得学习。

在设计沟通方面，很多人觉得建筑师在设计中起主导作用，造成其他专业设计的被动，其实只要搭建一个沟通的平台，各专业配合团队更早的介入，这种问题可以很好解决。比如我们此次邀请了做过柏林大剧院舞台设计的德国昆克国际咨询有限公司，从早期就开始和崔愷院士进行沟通，他需要一种怎样的支持，可上升的舞台需要多大面积，建筑师完全理解并全力配合。其他专业团队在这个项目中也都很早就开始介入，从设计初期就与建筑师保持了良好的沟通。

新印象：文化艺术中心最值得借鉴的地方在哪里？

周继春：很多地方的文化艺术中心造好了，可是却处于严重依赖财政补贴的境地，为什么？这值得我们思考。文化建筑不止要设计得合理，良性的运营模式更为重要，为此，我们决定和保利合作，因为他们有与国际各大演出团体的长期合作和管理的经验，并且我们被纳入到它的演出计划体系后，大大降低了邀请国际知名演出团体的成本。现在，我们每月的演出都有一个公益专场，保利在这方面也愿意反惠于民。5-10元的票价让普通市民可以欣赏传统的昆曲剧目，欣赏省昆剧院的演出。并且专门安排让市民了解如何欣赏昆曲的讲座。

如今，这里的新老昆山人可以看到爱丁堡艺术节上的节目，看到维也纳交响乐团演出的交响乐，看到著名的芭蕾曲目《天鹅湖》，这都是前所未有的视听享受，而且很多演出供不应求。事实证明当时市领导的决定还是深谋远虑的，我们的规划还是应该更有预见性，文化消费能力不是一蹴而就产生的，它也需要适当地加以引导。而且我们正陆续与上海、苏州的昆曲院开展合作，希望把诞生于昆山的昆曲“请回来”。目前已经开始这样做了。

新印象：您认为文化艺术中心带给了这个城市什么？对昆山市民的影响有哪些？

周继春：昆山市经济高速发展，如果它是一个国家的话，它的富庶程度已经不亚于新加坡。但是它的未来该走向哪里？这个城市的特色是什么？市民的生活方式与别的地方都有什么不同？此前我的妻子曾对我说：我不喜欢这座城市近些年的变化，我熟悉的河流不见了，我熟悉的树木不见了，渐渐地，我熟悉的环境与城镇气息也不见了。如果不是因为有家在这里，我完全可能哪天离开这个城市。这是一个值得每个城市里居住和生活的人深思的问题。

一个城市如果仅仅因为发展扩张而不尊重历史与传统，居住在这里的人们也会缺乏对它的认同感。这种城市居住者与过去的一切之间失去相互依存的关系就像某个人失去记忆一样，其后果不仅是严重的，在未来也是难以挽回的。因此我在主导这个项目计划的同时，也是在实现为昆山市和包括我自己在内的市民们提供一个文化载体，一个可以共同参与的大舞台，一块可以传播艺术和繁衍传统的土壤。并让新、老昆山人从精神上对城市有更多的认同感。同时，文化艺术中心的选址坐落在昆山西部新区，作为昆山西部副中心的有力补充，与运动馆，体育场有机地结合在一起，因而与森林公园连成一片，使这里真正成为一个完整的，多层面的丰富多样的市民活动中心。这无疑为西部副中心创造了新的吸引力，使片区产生连续的、值得关注的热点，同时将很好地刺激这片区域的良性发展。更重要的是它也适时地起到了以文化濡染昆山城市精神，在深厚的精神层面的精神层面振兴城市发展的作用。

New Impression: What are the different views of various parties in the preparation stage of Cultural and Arts Center project?

Zhou Jichun: the original plan for this project was to rebuild a theater based on the existing situations, however, when we invited the person in charge of the Shanghai Grand Theatre and the relevant experts of ECADI to give argumentation, their opinion is: Kunshan is not Suzhou or Shanghai, so it is not necessary to build a Grand Theatre, and a cultural center is sufficient. But the leaders of Kunshan City held different ideas, believing that Kunshan City is home of Kunqu Opera and enjoys rich spiritual and cultural heritage. And there is a foundation for the public to appreciate high art. So the final decision was: the Culture and Art Center not only functions as a theater and conference center, but also contains a large theater for Kunqu performance.

New Impression: In which stage did the Kunshan City Construction Investment & Development Co.,Ltd. involve in the project? What' s your thinking and plan for the preparation of Culture and Art Center? How did the self-positioning and design teams of the Culture and Art Center form?

Zhou Jichun: when the Kunshan City Construction Investment & Development Co.,Ltd. took over the project, the design bidding of the Art Center had been completed. I had my personal concern: although finanically Kunshan is capable of doing the project, there are too many unknowns. On which track will the project run? What kind of appearance will it show? An effective control must be carried out, otherwise the construction of a high-quality work cannot be ensured. To meet this end, we have adopted a series of measures. Firstly, we asked our company team to do a thorough investigation on the current status of the theaters of some representative metropolis, and they made a field understanding of the theater design and operations of Shanghai, Suzhou, Zhongshan, Hangzhou, Wuhan, Chongqing, Shenzhen, Beijing and other cities with questions and purposes. According to findings of the study, we have mastered that what state we should have to face it, what experience is worthy of our recommendation, what issues we should avoid, and what professional teams should be formed to realize the design, management design and achieve an orderly operation finally. Ultimately, we want it to run on a controlled and benign orbit. Here, I think that although Zhongshan City Art Center was built under the leading of cultural arts community, the operating model of it is very good. In my opinion, other theaters have a lot of flaws, and a lot of things are not worthy of imitation.

In the aspect of design communication, many people hold that the architect takes the leading role in the design, causing other professional design to be passive; in fact, if the professional teams get involved in a project earlier, this problem can be solved with no doubt. For example, we have invited Germany Kunke International Consulting Ltd. who conducted the stage design of Berlin Grand Theatre to communicate with academician Cui Kai in the early stage, which is concerned about what kind of support he needs, how much space the rising stage needs, and the architect understands and cooperated completely. Other professional teams also have been involved in this project from the early days, and they maintain a good communication with the architect from the beginning of design.

New Impression: What' s the aspect of the Culture and Art Center that is most worthy of learning?

Zhou Jichun: Some culture and art center has been completed, but it is heavily dependent on financial subsidies, why? It's worthy of thinking about. Cultural buildings should not be only designed reasonable, and a benign mode of operation is more important. To this end, we have decided to cooperate with Poly, because they have experience in cooperating with and managing major international performing groups for a long term, in addition, being incorporated into its performance plan system can significantly reduce the costs of inviting internationally renowned performing groups.At present, we have a special public welfare performance for our monthly performances, and Poly is also willing to feed benefit back to the public. The ticket pricing of 5-10 Yuan can make ordinary citizens enjoy a traditional opera repertoire and enjoy the performances of provincial Kunqu Opera. What's more, we also arrange lectures for the citizens to understand how to appreciate Kunqu.

At present, the new and old Kunshan citizens can enjoy the shows on Edinburgh Festival, symphony performed by Werner Symphony Orchestra and the famous ballet repertoire Swan Lake, which are unprecedented visual enjoyment; moreover, many performances are actually in short supply. Facts prove that the decisions of city leaders is thoughtful, and our planning should be more proactive, while cultural consumption capacity is not produced overnight, and it also needs to be guided properly. We begin to cooperate with Kunqu Opera Institutes of Shanghai and Suzhou, hoping to invite the Kunqu Opera born in Kunshan "back". We have already begun to do it.

New Impression: Do you think what the Culture and Art Center brings to the city? What effect does it exert on Kunshan citizens?

Zhou Jichun: Kunshan City meets a rapid economic development, and its affluence level has not been less than Singapore if it were a country. But where should it go in future? What are the characteristics of this city? Does the citizens' lifestyle differ from that of other places? My wife told me: I do not like the recent changes of this city, because the rivers and trees I'm familiar with disappeared, and the environment and urban flavor is gone gradually. If not for a home here, I might leave the city someday. This is a question worth us thinking about, including the people who are living in each city. If a city does not respect history and tradition any more just due to development and expansion, the people who live there will lack sense of identity to it. Losing interdependent relationship between the city dwellers and all the past is like a person losing his memory, so the consequence is not only serious, but also very difficult to restore in the future. Therefore, while I was leading the project plan, I was also trying to provide a carrier of culture, a big stage we can participate in, and a piece of soil that can spread arts and produce traditions for Kunsan city and citizens including me, and make the new and old Kunshan citizens have more sense of identity to the city. Meanwhile, the Culture and Art Center is locate in the western new district of Kunshan; as a strong complement to the western sub-center, thus, being organically together with sports hall, and stadium, it forms an unbroken expansion with the forest park, making this place a truly complete, multifaceted and various public activity center. This undoubtedly creates a new attraction for the western sub-center, and makes the areas have continuous hotspots which are worth focusing on; meanwhile, it can greatly stimulate the benign development of this area. More importantly, it also plays a plays a timely role in influencing Kunshan city spirit with culture and revitalizing the urban development at a deep spiritual level.

姚伟宏，昆山市文化广电新闻出版局局长、昆山市广播电视台台长

YaoWeihong, director general of Kunshan Municipal Culture, Radio, Press and Publication Bureau, head of Kunshan Broadcasting and TV Station

Interview with Yao Weihong

Social Value of Venues for Art

艺术场所的社会价值·姚伟宏

新印象：文化艺术中心的建成对昆山的重大意义有哪些？

姚伟宏：文化艺术中心的建成满足了当地市民多层次的文化需求。昆山的非物质文化遗产非常丰富，其文化资源不仅历史悠久而且广博精深，因此可以说对传统文化和对高雅艺术的欣赏，这里有着天然的土壤。以前欣赏这类演出，昆山市民需要到上海、苏州等地方，钱没少花，还很不方便。基于这样的基础和现实状况，随着昆山文化艺术中心建成，我们与保利文化集团股份有限公司成立了一个合资公司，借助保利的资源和管理经验，我们每年引进100场高端艺术节目，其中30场作为面向市民开放的节目，价位低廉；另外70场之中，有40场则由世界著名的剧团演绎国际知名的剧目。因此极大地丰富了这里的人们对高雅艺术的欣赏需求，目前上座率达90%左右，由此可以看出昆山市场潜力很大，文化艺术需求很高。事实也证明昆山文化艺术中心的市场化运作是深谋远虑的、超前的。同时，剧院配套设施与相关专业设计也是一流的，得到了来此巡回演出的维尔纳、俄罗斯等著名艺术演出团体的高度评价。

新印象：昆山文化艺术中心与其他文化场所最明显的不同在哪方面？作为昆曲的发源地，在昆山，通过文化艺术中心的建成，昆曲会得到哪些支持？

姚伟宏：昆山文化艺术中心有很多自身的特色，但我觉得其最突出的特色在于：它是一个文化惠民工程，在这里体现了什么叫高贵不贵，体现了一个高雅艺术场所的社会效益与现实价值。它的剧院年均票价200元左右，在演出季里，每个月都有一场惠民演出，花5元至10元成为会员后可以凭此看一场演出，基本属于无偿票价。同时它未来将作为百姓的大舞台，每年要安排一批大型群众性演出。而昆山文化艺术中心的多功能剧场将会成为老百姓的昆曲大舞台。我们正陆续邀请江苏省昆剧团，苏州昆曲团来此进行传统剧目的演出。并针对目前昆曲的发展现状和观众的需求，专门请昆曲界的名人、专家为喜欢昆曲的市民们讲解昆曲艺术欣赏知识，让昆曲的艺术魅力更广泛的为人们所熟知。

新印象：未来文化艺术中心的计划与定位包括什么？具体到对昆曲来讲，现实和未来意义有哪些？

姚伟宏：在昆山文化艺术中心的保利大剧院里，其未来的演出计划我们准备每年增加10场以上。另外一项重要的举措则是根据目前戏剧发展的严峻现实出发而产生的计划，目前昆山没有自己固定的昆曲演出团体，而以目前戏剧市场化的发展趋势来看，成立一个完全靠财政拨款作为来源的剧团是不符合现实的。

由此，我们正在和保利文化集团商讨，以项目合作的形式，通过与著名昆曲演员签约、加盟等形式，力图探索一种新的模式把昆曲“请回来”。

昆山作为昆曲的故乡希望能打造自己的演出团队。这可能是个坎坷艰辛的过程，但是作为昆曲的发源地，我们感到有着一种义不容辞的责任。

New Impression: What significances do the establishment of Culture and Art Center have on Kunshan?

Yao Weihong: The establishment of Culture and Art Center meets the multi-level cultural needs of local residents. Kunshan has rich intangible cultural heritages. Its cultural resources are time-honored, extensive and profound. Therefore, there is natural soil for appreciating traditional culture and elegant art. In the past, Kunshan's citizens needed to go to Shanghai, Suzhou and other places to enjoy this kind of performances, which is expensive and inconvenient. In view of this basis and realistic conditions, with the establishment of Kunshan Culture and Art Center, we set up a joint venture with Poly Culture Group Corporation Limited. With the help of Poly' s resources and managerial experiences, we introduce 100 high-end art programs every year, 30 of which are at low prices as programs open to citizens. In the remaining 70 programs, 40 of them are internationally renowned plays performed by famous troupes in the world. Consequently, it greatly enriches the appreciation demands of people here on high art. At present, with an attendance of about 90%, it can be observed that Kunshan has a large market potential and high culture and art demands. It is also proved that Kunshan Culture and Art Center's marketing operation is foresighted and advanced. Meanwhile, the supporting facilities and relevant professional designs are top-ranking as well, being highly appraised by Werner, Russia and other famous artistic performance groups.

New Impression: What aspect is the most obvious difference between Kunshan Culture and Art Center and other venues for art in? As the birthplace of Kunqu opera, through the establishment of Culture and Art Center, what supports will Kunqu opera get in Kunshan?

Yao Weihong: Though Kunshan Culture and Art Center has various features, the most prominent one, in my opinions, is that it is a people-benefit culture project where noble but not expensive art is reflected and the social benefits and realistic values of an elegant venue for art is embodied. The annual average ticket price of the theatre is about 200 RMB. During the performance season, there will be a avdience-benefit performance every month, and people can enjoy a performance after becoming a member with 5-10 RMB, which belongs to free tickets basically. At the same time, it will schedule a batch of large mass performances every year as a grand stage for people in the future. The multifunctional theatre in Kunshan Culture and Art Center will become a grand stage of Kunqu opera for people. We are inviting Jiangsu Kunqu Opera Theatre and Suzhou Kunqu Opera Troupe to perform traditional theatrical pieces. Aiming at the development status of Kunqu opera and the demands of the audiences at present, celebrities and experts in the circle of Kunqu opera are invited specially to explain the appreciation knowledge of Kunqu opera art to citizens making the artistic charm of Kunqu opera become much more well-known.

New Impression: What are included in the future plan and positioning of Culture and Art Center? As for Kunqu opera specifically, what are the realistic and future meanings?

Yao Weihong: In the Poly theatre of Kunshan Culture and Art Center, we plan to add more than 10 performances every year. Another important action is that we will plan according to the rigorous reality confronted by operas when developing at the moment. Currently, there is not permanent Kunqu opera performance group in Kunshan, while seen from the development tendency of opera marketization at present, it is not realistic to establish a troupe funded by financial allocation completely. Therefore, we are discussing with Poly Culture Group Coorporation Limited, in the form of project cooperation, through signing contracts with famous Kunqu opera performers, franchisees and other ways, to explore a new mode to "invite" Kunqu opera. Kunshan, as the birthplace of Kunqu opera, hopes to create its own performance teams. Although this may be a difficult process, we feel that it is our bounden duty.

韩明，昆山市保利大剧院管理有限公司 总经理

Han Ming, general manager of Kunshan Poly Grand Theatre Management Co., Ltd.

Interview with Han Ming

Carrying Forward Tradition in Sound Operation

在良性运营中弘扬传统·韩明

新印象：你们在考察这个项目时，感觉文化艺术中心与其他保利介入的剧院相比，特点都有哪些？感觉如何？您如何评价昆山艺术中心的建成？

韩明：个人感觉昆山文化艺术中心与其他剧院相比，虽然不是最大，甚至不具有夺人眼球的那种强势，但它的设计风格很独特。非常能够深入体现当地标志性特色，尤其是崔愷大师设计中体现的与水乡、与昆曲水袖的意念相契合之处。反映了设计师对这个城市特色的精准把握。包括外部幕墙设计的变化手法，很生动。具体到室内的设计以及各种配备，与其他院线相比，都属于比较高端的。

新印象：昆山目前已开展的剧目有哪些？与观众互动和反馈怎么样？

韩明：项目建成后，在昆山保利大剧院管理有限公司管理下，到现在已成功安排了20多场演出。保利的剧目非常丰富，涵盖了各种演出形式，包括话剧、歌舞等等。从爱尔兰踢踏舞到东方歌舞团的演出、音乐会、个人演唱会等等，社会反响很乐观，昆山市民的评价也很高。市民也因此对这里的演出有了更多的期待，我们也要随时了解他们的需求，不断满足这种期待。

新印象：目前运营效果如何？与业主的合作建立在一种什么基础上和模式？

韩明：从我们的介入目的是要促成业主与我们的客户的合作达到包场、团购、冠名这些经营性活动的实现以完善的体系支持各系列演出剧目成功上演。我们在开展业务，促成各种商务合作的时候，客户会希望充分利用这高雅的艺术表演场所的大堂、广场等各种广告位。但是我们同时也要考虑业主对整个艺术中心的文化品位与自我定位的要求，如杨坤演唱会期间，其中一家地产赞助商在商务洽谈过程中，试图把地产楼盘的宣传以展览的形式放到艺术中心里面，业主方对此表示对昆山文化艺术中心，这样一个传播艺术与精神文化的地方，这样的商业模式会有损整体形象。最后我们通过其他增值服务等合作方式与赞助商沟通，地产开发商不再坚持原有方案。双方都做到了满意，由此可见随着客户自身素质的提高，对值得坚持的原则还应保持自己的底线。合理的他们也是可以接受的。目前保利在实现了把国内外高雅艺术全面带给昆山观众的初期目标的同时，整个运营效果相当不错。

新印象：根据以往保利集团管理剧院的丰富经验，昆山大剧院将被纳入一种怎样的模式？对未来发展有哪些计划和新的定位思考？

韩明：保利的模式全年分为四大演出季，分为新春演出季、戏曲舞蹈演出季、打开艺术之门、保利流行风。随着剧目的规整，根据不同的院线有所调整，结合当地情况引进国际一流演出。下一阶段我们即将开展“打开艺术之门”的计划，这是针对少儿的普及活动。是具有公益意义的会员的运营形式，它最高的票价也就是100元。基本保持30-60元的票价。另外，每个月第四周的周日定为市民的免费参观日。同时每到这一天下午都引进一场省昆院的演出，作为昆曲艺术普及活动。每人五元钱。通过免费办理的会员卡的方式，持有会员卡的观众五元就可享受一场演出。我们从落户昆山就有这样的计划，从保利进驻昆山之日起，就要发扬本土的文化。这样昆山市政府和居民才会认为引进保利是真正意义地满足了昆山精神文化的需求和发扬文化产业的需求。为此保利的管理层也花费了不少心思，全力要把昆曲做好。我们的目标是既要把传统艺术发扬好，又要让它纳入商业良性循环的轨道。

New Impression: When investigating this project, what characteristics do you feel comparing Culture and Art Center to other theatres Poly involved in? How do you evaluate the establishment of Kunshan Culture and Art Center?

Han Ming: Comparing to other theaters, although Kunshan Culture and Art Center is not the biggest and does not possess the eye-catching strength, it has a unique design style and can reflect the local landmark characteristics deeply. The design of Master Cui Kai reflects some points agreeing with the ideas of watertown and actor's long sleeves in Kunqu opera, embodying the designer's precise understanding on the characteristics of this city. It is vivid including the changes in the design of external curtain wall. Comparing with other theaters, it is high-end as for interior design and all kinds of equipment.

New Impression: What plays have Kunshan developed at present? How about the interaction with the audience and feedback?

Han Ming: After the completion of this project, under the management of Kunshan Poly Grand Theatre Management Co., Ltd., more than 20 performances have been arranged successfully by this time. Poly has rich plays, covering all kinds of performing forms, for example, drama, song and dance, etc. The Irish step dance, performances of China Song and Dance Ensemble, concerts, solo concerts, etc. all have positive social response and are highly appraised by Kunshan residents. For this reason, residents expect much more on the performances in here. We also need to understand their demands at any time and fulfill this expectation continuously.

New Impression: What's the current sitation of operation? What basis and pattern do the cooperation with the owner establish on?

Han Ming: Our intervention is aimed to promote the cooperation between the owner and our clients, implement commercial activities including block booking, group buying and title sponsoring and support the performance of all series of plays with perfect system. When we conduct business and promote all kinds of commercial cooperation, the clients would like to make the most of the lobby, square and other advertisement sites in the elegant artistic performance venue. However, we need to consider the owner's requirements on the cultural taste and self-positioning of the whole arts center. For example, during Yang Kun's concert, one real estate sponsor tried to put the advertisements of real estate and houses in the Art Center in the form of exhibition in the process of business negotiations, while the owner argued that on Kunshan Culture and Art Center, a place disseminating art and spiritual culture, this kind of business model would spoil the overall image. At last, we communicated with the sponsor through other value-added services and other ways of cooperation, the property developer did not stick to the original scheme, and both parties were satisfied. This shows that with the improvement of clients' quality, we should keep our bottom line for worthwhile principles. They can also accept the reasonable things. At present, while Poly achieves the preliminary goal of bringing the elegant art at home and abroad to the audience in Kunshan, its whole operation effect is pretty good.

New Impression: According to the rich experience of Kunshan Poly Grand Theatre Management Co., Ltd. in the past, which kind of model will Kunshan Grand Theatre fit into? What plans and new reflection on positioning does it have on future development?

Han Ming: Poly's model in the whole year includes four performance seasons, namely, new spring performance season, traditional opera and dance performance season, opening the door to art and Poly fashion. With the structuring of plays, it adjusts according to different theatre chains, and combines with local conditions to introduce world-leading performances. In the next stage, we will develop the plan of "opening the door to art", a popularization activity for children and a member operation form with social significance. Its ticket price remains in 30-60 RMB generally, and would reach the highest of 100 RMB. Moreover, the fourth Sunday every month is the free visiting day for citizens. Meanwhile, in the afternoon of this day, a performance by the provincial Kunqu opera theatre will be introduced as a popularization activity of Kunqu opera, 5 RMB per person. Through the way of getting membership card freely, the audiences holding membership cards can enjoy a performance with 5 RMB. Since our settlement in Kunshan, we have such a plan that from the day Poly entering Kunshan, it should promote local culture. Thus, Kunshan Municipal Government and residents can deem that the introduction of Poly satisfies the needs of Kunshan spiritual culture and developing culture industry in the real sense. For this purpose, Poly's management also thinks carefully and spares no effort to develop Kunqu opera. Our goals are to carry forward traditional art and bring it into the orbit of commercial virtuous cycle.

项目信息
Project Information

昆山文化艺术中心
Kunshan Culture and Art Center
中国，江苏昆山
Kunshan, Jiangsu Province, China

业主 Client
昆山城市建设投资发展有限公司
甲方项目总协调
General project coordinator of clients
范晓玲
建筑设计 Design architects
中国建筑设计研究院
主持建筑师 Principle
崔愷
项目总监 Project director
马海
项目团队 Project team
方案设计 Plan design
崔愷，李斌，何理建(德)，俞弢，关飞
设计主持 Design manager
崔愷，何咏梅
建筑 Construction
李斌，张玉明，颜朝昱，于玢，曾瑞
总图 General plan
白红卫
结构 Structure
朱炳寅，王奇，张根俞
给排水 Water supply and drainage
侯远见，高峰
暖通空调 HVAC
金跃，贺舒
电气电讯 Electric telegraphen
陈琪，王烈，程培新，崔振辉，胡桃
建筑预算 Construction budget
贺玉萍
室内方案设计 Interior design
张晔，张明杰，纪岩，饶劢，盛燕
邸士武，马盟雪，江鹏，郭林，
韩文文，刘烨，张然，王墨涵，李毅
室内施工图设计
Interior construction drawing design
深圳洪涛装饰股份有限公司
可持续节能设计
Sustainable energy conservation design
中国建筑科学研究院上海分院
主持建筑师 Principle
邵文晞
舞台设备咨询与设计
Stage equipments consultant and design
Kunkel Consulting International GmbH
声学顾问
Acoustics consultant
张三明
色彩顾问 Color consultant
Alain Bony
幕墙工程设计与咨询
Curtain wall design and consultant
上海和甲幕墙设计咨询有限公司
幕墙工程施工图设计
Curtain wall project construction drawing design
苏州苏明装饰有限公司
照明设计顾问 Lighting design consultant
Leoxdesign partnership(ShangHai)
照明设计 Lighting design
Leoxdesign partnership(ShangHai)，北京良业照明技术有限公司
景观设计 Landscape design
香港周保华设计有限公司
机电顾问 Mechanical consultant
迈进机电工程顾问有限公司
人防工程施工图设计
Civil air defense works construction drawing design
苏州市天地民防建筑设计研究院有限公司
剧院设计咨询 Theatre design consultant
北京保利剧院管理有限公司
影院设计咨询 Cinema design consultant
上海联和电影院线有限责任公司
总建筑面积 Overall floor area
80 551.8 m^2
功能 Program
大剧院 Theatre
32 845m^2，
会议中心 Conference center
10 173m^2
影视中心 Media center
12 193m^2
辅助设施 Auxiliary facilities
16 044m^2
景观 Landscape
53 865m^2
绿化率 Greening rate
45%
容积率 Plot ratio
0.43
施工阶段 Construction phase
2009 - 2012

索引
Index

出 品 人：
叶春曦
主　　编：
于 冰
主编助理：
李 华
执行编辑：
梁 万
英文编辑：
王僖婷
美术编辑：
李 渔　黄 洁　逄 苹
摄　　影：
胡文杰

Producer:
Isa Ye
Chief Editor:
Yu Bing
Assistant to Chief Editor:
Li Hua
Executive Editor:
Liang Wan
English Editor:
Wang Xiting
Art Editor:
Li Yu / Huang Jie / Pang Ping
Photography:
Hu Wenjie

编者对以下提供支持的单位和个人深表感谢：
昆山市保利大剧院管理有限公司、中天建设集团有限公司、昆山市文化发展研究中心以及卢涛、张娜、陈益、楼良科。

图书在版编目(CIP)数据

生长的文化建筑 昆山文化艺术中心全解析：汉英对照 / 于冰主编. —北京：中国建筑工业出版社，2013.9
（新印象）
ISBN 978-7-112-15873-7

Ⅰ.①生… Ⅱ.①于… Ⅲ.①文化建筑–建筑设计–昆山市–汉、英 Ⅳ.①TU242

中国版本图书馆CIP数据核字(2013)第221442号

责任编辑：徐 纺
责任校对：陈晶晶

新印象
生长的文化建筑 昆山文化艺术中心全解析
于冰 策划&主编
*
中国建筑工业出版社出版、发行（北京西郊百万庄）
各地新华书店、建筑书店经销
北京建宏印刷有限公司制版印刷
*
开本：880×1230毫米 1/8 印张：20¼ 字数：325千字
2013年9月第一版 2013年9月第一次印刷
定价：200.00元
ISBN 978-7-112-15873-7
（24649）